OECD PROCEEDINGS

DISPOSAL OF RADIOACTIVE WASTE

Field Tracer Experiments: Role in the Prediction of Radionuclide Migration

Synthesis and Proceeding
of an NEA/EC GEOTRAP Workshop
hosted by the Gesellschaft für Anlagen- und Reaktorsicherheit (GRS)
Cologne, Germany, 28-30 August 1996

**A workshop organised in the framework of the NEA Project
on Radionuclide Migration in Geologic, Heterogeneous Media
(GEOTRAP)**

PUBLISHER'S NOTE
The following texts are published in their original form to permit faster distribution at a lower cost

NUCLEAR ENERGY AGENCY
ORGANISATION FOR ECONOMIC CO-OPERATION AND DEVELOPMENT

ORGANISATION FOR ECONOMIC CO-OPERATION AND DEVELOPMENT

Pursuant to Article 1 of the Convention signed in Paris on 14th December 1960, and which came into force on 30th September 1961, the Organisation for Economic Co-operation and Development (OECD) shall promote policies designed:

- to achieve the highest sustainable economic growth and employment and a rising standard of living in Member countries, while maintaining financial stability, and thus to contribute to the development of the world economy;
- to contribute to sound economic expansion in Member as well as non-member countries in the process of economic development; and
- to contribute to the expansion of world trade on a multilateral, non-discriminatory basis in accordance with international obligations.

The original Member countries of the OECD are Austria, Belgium, Canada, Denmark, France, Germany, Greece, Iceland, Ireland, Italy, Luxembourg, the Netherlands, Norway, Portugal, Spain, Sweden, Switzerland, Turkey, the United Kingdom and the United States. The following countries became Members subsequently through accession at the dates indicated hereafter: Japan (28th April 1964), Finland (28th January 1969), Australia (7th June 1971), New Zealand (29th May 1973), Mexico (18th May 1994), the Czech Republic (21st December 1995), Hungary (7th May 1996), Poland (22nd November 1996) and the Republic of Korea (12th December 1996). The Commission of the European Communities takes part in the work of the OECD (Article 13 of the OECD Convention).

NUCLEAR ENERGY AGENCY

The OECD Nuclear Energy Agency (NEA) was established on 1st February 1958 under the name of the OEEC European Nuclear Energy Agency. It received its present designation on 20th April 1972, when Japan became its first non-European full Member. NEA membership today consists of all OECD Member countries, except New Zealand and Poland. The Commission of the European Communities takes part in the work of the Agency.

The primary objective of the NEA is to promote co-operation among the governments of its participating countries in furthering the development of nuclear power as a safe, environmentally acceptable and economic energy source.

This is achieved by:

- *encouraging harmonization of national regulatory policies and practices, with particular reference to the safety of nuclear installations, protection of man against ionising radiation and preservation of the environment, radioactive waste management, and nuclear third party liability and insurance;*
- *assessing the contribution of nuclear power to the overall energy supply by keeping under review the technical and economic aspects of nuclear power growth and forecasting demand and supply for the different phases of the nuclear fuel cycle;*
- *developing exchanges of scientific and technical information particularly through participation in common services;*
- *setting up international research and development programmes and joint undertakings.*

In these and related tasks, the NEA works in close collaboration with the International Atomic Energy Agency in Vienna, with which it has concluded a Co-operation Agreement, as well as with other international organisations in the nuclear field.

FOREWORD

GEOTRAP, the OECD/NEA Project on Radionuclide Migration in Geologic, Heterogeneous Media, is devoted to the exchange of information and in-depth discussions on present approaches to acquiring field data, and testing and modelling flow and transport of radionuclides in actual geologic formations for the purpose of site evaluation, and safety assessment of deep repository systems. The project is articulated in a series of structured, forum-style workshops.

The first GEOTRAP workshop, *"Field Tracer Experiments: Role in the Prediction of Radionuclide Migration"*, was held in Cologne (Germany) on 28-30 August 1996. It was co-organised with the Directorate General XII (Science, Research and Development) of the European Commission, and was hosted by the *Gesellschaft Für Anlagen- und Reaktorsicherheit, GRS, mbH* (German Company for Reactor Safety).

The workshop was aimed at providing a structured forum whereby implementors, regulators and scientists could interact, contribute to the advancement of the state of the art in this area, discuss the approaches and rationale of past, current and planned tests, and assess the results and uses of past experiments.

In addition to oral and poster presentations, the workshop consisted of focused discussions within four working groups.

The technical presentations gave an overview of on-going and planned work in the study of radionuclide transport phenomena and the characterisation of relevant properties of the geologic media. Discussions took place on the extent to which it is possible to resolve migration problems using field tracers experiments, and all participants were asked to define the role of tracer tests in the safety assessment of deep radioactive waste repositories.

This publication includes a synthesis of the workshop that reflects the materials that were presented, the discussions that took place and the conclusions drawn, notably during the working group sessions. The publication also reproduces the papers presented at the workshop. The opinions, conclusions and recommendations expressed are those of the authors only, and do not necessarily reflect the view of any OECD Member country or international organisation. This report is published on the responsibility of the Secretary General of the OECD.

ACKNOWLEDGEMENTS

On behalf of all participants, the NEA wishes to express its gratitude to the *Gesellschaft für Anlagen- und Reaktorsicherheit, GRS, mbH* (Germany) which hosted the workshop at its Cologne premises, and to the Fuel Cycle and Radioactive Waste Unit (F5) of the Directorate General XII (Science, Research and Development) of the European Commission which co-organised the workshop.

The success of the workshop was due to:

- the members of the workshop Programme Committee: Peter Bogorinski (GRS, Germany), Russell Alexander (NAGRA, Switzerland), Juhani Vira (POSIVA, Finland), Geert Volckaert (SCK/CEN, Belgium), Henning von Maravic (European Commission), Philippe Lalieux (OECD/NEA), and Claudio Pescatore (OECD/NEA);
- the NEA consultants who helped conduct the numerous discussions: Paul Smith (Safety Assessment Management Ltd, United Kingdom), Aimo Hautojärvi (VTT Energy, Finland) and Mike Heath (Earth Resources Centre, Exeter University, United Kingdom), and who also helped the Secretariat in drafting the synthesis; and
- the speakers, the posters' authors and other participants;

all of whom deserve thanks for their active and constructive contribution.

The Chairmen of the sponsoring NEA groups, Alan Hooper (Co-ordinating Group on Site Evaluation and Design of Experiments) and Piet Zuidema (Performance Assessment Advisory Group), made a number of valuable comments and contributions in reviewing the synthesis; their reviews have been addressed in the present publication.

Claudio Pescatore and Philippe Lalieux from the Radiation Protection and Waste Management Division of the OECD/NEA are responsible for the GEOTRAP project's scientific secretariat.

TABLE OF CONTENTS

PART A: SYNTHESIS OF THE WORKSHOP

PART B: WORKSHOP PROCEEDINGS

SESSION I

General Overview
Chairmen: C. Pescatore (NEA) and H. von Maravic (EC)

SESSION II

Rationale Behind Field Tracer Experiments
Chairmen: M. Heath (Earth Resources Centre, United Kingdom)
and P. Bogorinski (GRS, Germany)

SESSION III

Test Cases: Design, Modelling and Interpretation
Chairmen: J. Vira (Posiva, Finland) and P. Lalieux (NEA)

SESSION IV

Aims and Design of Planned Field Tracer Experiments
Chairmen: W.R. Alexander (Univ. Berne, Switzerland) and G. Volckaert (SCK-CEN, Belgium)

POSTER SESSION

PART A

SYNTHESIS OF THE WORKSHOP

EXECUTIVE SUMMARY

Several past, current and planned field tracer experiments were described and discussed in the course of the workshop. They cover various potential repository host rocks from soft argillaceous media to hard, fractured crystalline basement; make use of a wide range of both sorbing and non-sorbing (conservative) tracers; and cover various types of geologic features at different spatial and temporal scales.

The workshop provided a broad perspective on the advantages and limitations of field tracer experiments, and a set of conclusions and recommendations that will be useful when designing future tests. The main conclusions of the workshop are as follows:

1. Field tracer experiments have a valuable role to play in building confidence in the identification of the processes relevant to transport, in the provision of parameter values that are required by transport models and in the definition of site-specific models.

2. A good characterisation of the flow system in the region of the test is desirable for a meaningful interpretation of tracer experiments and, in particular, for reducing the degree of non-uniqueness.

3. Although the interpretation of the results of tracer experiments can be non-unique in terms of the operating processes, particularly where the structure of the system is incompletely characterised, no new processes, outside the scope of the current models, need to be invoked to rationalise the experimental results. This contributes to confidence that the processes relevant to geosphere transport have been identified. The structural complexity of natural systems remains a significant source of uncertainty, both in the interpretation of tracer experiments and in the modelling of the performance of deep repository systems.

4. The relative importance of the operating processes, as well as the complexity of structure, varies among different geologic media. This has strong implications for the type of tests to be performed, the type of information that can be obtained and its uses for performance assessment.

5. Tracer tests are most likely to be valuable when planned by a multidisciplinary team, including experimentalists, hydrogeological modellers and performance assessment specialists.

6. In the assessment of the geological barrier of a repository system, integration of tracer tests with other types of studies (e.g. paleohydrogeology) is seen to be crucial, as are the testing of alternative hypotheses and the identification of features of repository host rock relevant to flow and transport. Field tracer experiments in isolation can never provide an adequate proof of the performance of the geosphere.

7. There are inherent limitations in the use of field tracer experiments in support of the assessment of the geosphere as a barrier to radionuclide migration:

- The experimental time and length scales and flow conditions that can never fully reproduce those relevant to performance assessment. The length scales that have been explored experimentally are, however, very relevant for the analysis of the transport properties of the near-field host rock.

- The practical difficulties in integrating these experiments with other studies.

- The limited transferability of data from one site to another.

In summary, field tracer experiments should continue to form part of the research needed for performance-assessment modelling as they provide important technical information, help build confidence in performance-assessment models, and provide much, sometimes unexpected, supporting information and understanding of radionuclide-migration mechanisms. Among the wider benefits to be derived from the continued use of field tracer experiments are also the development of interdisciplinary teams and improvement of public confidence in disposal options, though it should be remembered that the results of field tracer experiments can be difficult to interpret and that such tests cannot provide proof of the performance of the geological barrier of any disposal system, but are rather a technique among others to be used in pursuit of a solution to the safe disposal of radioactive waste.

1. INTRODUCTION

1.1 RATIONALE AND STRUCTURE OF THE FIRST GEOTRAP WORKSHOP

The use of field tracers is a most prominent approach to study flow distribution, characterise potential flow paths, test different conceptualisations (both of flow and transport) and estimate transit times at selected sites and at different scales. The experience to date from field tracer experiments demonstrates the complexity of the techniques used. This led to the decision, within the framework of GEOTRAP, to hold the first workshop on the rationale and planning of tracer experiments, keeping in mind the needs of site characterisation and performance assessment.

The goals of the workshop were:

- To provide a forum whereby implementers, regulators and scientists can interact in a structured fashion.

- To learn about and contribute to the advancement of the state of the art in the area of field tracer experiments in order to build confidence in the modelling of radionuclide transport in geologic, heterogeneous media.

- To comment on the approaches used by different programmes.

- To discuss the rationale, objectives and strategies for past and planned tests.

- To evaluate the uses and results of field tracer tests in the light of alternative testing methodologies.

- To assess the results of past and current tests and their uses/relevance for site characterisation and performance assessment purposes.

- To compare "generic" tests with site-specific tests.

A final aim of the workshop was the preparation of this synthesis, which reviews and summarises the lessons learned at the workshop, putting them into perspective within the scope of the GEOTRAP Project and the state of the art in this field.

The workshop was introduced by three overview papers in order to provide the audience with a common background for the planned discussions. The three sessions addressed the rationale behind tracer tests, presented several test cases and discussed the aims of planned experiments, respectively. In addition, a poster session dealt with more specific, technical details. A key part of the workshop consisted of focused discussions within small working groups, on specific themes. The outcomes of the working groups provided the basis for a plenary, concluding discussion. For each session and for each working group, the Programme Committee had established a series of key questions to be addressed. This proved to be a very effective way of focusing the discussions and reaching practical conclusions.

1.2   STRUCTURE OF THE SYNTHESIS

The workshop is synthesised at three different levels by providing:

1. An overview of the workshop achievements, which includes an assessment as to how each goal specified for the workshop was met, the general conclusions as well as the inherent limitations of the tests and recommendations regarding future work (Chapter 2).

2. Synoptic tables of the key features of the various field tracer experiments that were described in the course of the presentations and discussions (Annex 1). These tables help set the context in which the achievements, general conclusions, limitations and recommendations should be viewed by providing an overview of the main current and planned tests, including their status and principal aims.

3. A detailed record of the workshop that compiles the answers to the key questions specified for the four workshop sessions, and reports the discussions and conclusions of the four working groups and of the final, concluding session (Annex 2).

2.   WORKSHOP ACHIEVEMENTS

2.1   ACHIEVEMENTS OF THE WORKSHOP GOALS

The extent to which the principal goals of the first GEOTRAP workshop, described in Chapter 1.1, were achieved can be summarised as follows:

- **To provide a forum whereby implementers, regulators and scientists can interact in a structured fashion.**

The workshop was attended by 40 delegates from 12 countries, with a range of experience including performance assessment, site characterisation, experimental techniques and modelling. Several implementing organisations and a few regulatory bodies were represented, as well as research laboratories and universities. The workshop comprised 14 technical presentations, each allocated a discussion period, and in-depth discussions by 4 working groups on a range of detailed topics. The technical presentations triggered discussion as to the extent to which it is possible to resolve migration problems using field tracer experiments, and particularly problems relevant to performance assessment; these discussions ultimately led to some constructive ideas for improvements of future experimental programmes. There was no special weighting of issues from the point of view of interest to implementors, regulators or scientists, but regulators were, to some extent, challenged to define the role of tracer tests in performance assessment.

- **To learn about and contribute to the advancement of the state of the art in the area of field tracer experiments in order to build confidence in the modelling of radionuclide transport in geologic, heterogeneous media.**

The technical presentations gave an overview of on-going work on the study of migration phenomena and the characterisation of properties of geologic media that are relevant to the transport of radionuclides. Planned and performed tracer experiments related to waste-management programmes in various countries were presented and discussed thoroughly in the workshop sessions. The experience gained, and the continuing need for improvement, were summarised in the conclusions drawn by the four working groups. The overall contribution to the field can be judged from the conclusions and recommendations of the workshop. The consensus among the participants was that the workshop was successful and useful, both to themselves and, more generally, to organisations working in the field of radioactive waste disposal.

- **To comment on the approaches used by different programmes.**

From the presentations and discussions of the workshop, it was apparent that the approaches used by different programmes (i.e. whether the tests are used to provide basic understanding of processes or to test and refine performance-assessment models) are dependent on the "simplicity" or "complexity" of the host rock under consideration (e.g. plastic clays vs. fractured rocks) and on the stage reached by the national waste management programme. There is also a contrast between programmes that emphasise the use of field tracer experiments to develop understanding of the transport of non-reactive tracers in systems of increasing spatial scale (and structural complexity) and those that emphasise the use of increasingly complex tracers in structurally relatively simple systems. The existence of experimental artefacts was identified as a problem area faced by all programmes. This, and other practical challenges, were addressed by Working Group 1 *(Practical Challenges)*.

- **To discuss rationale, objectives and strategies for past and planned tests.**

The changing rationale of field tracer experiments in recent years was noted. Such experiments were previously considered indispensable for "overall validation" of radionuclide transport models. These expectations are now regarded as having been over-ambitious. Field tracer experiments are now better focused and generally adopt a strategy of beginning with simple systems (structural simplicity, non-sorbing tracers), before progressing to more complex systems once the simple systems are fully understood. The desirability of participation by different parties involved in site characterisation and performance assessment was noted, as was regulatory interest (although in no country are field tracer experiments mentioned in regulatory guidelines). Current expectations from field tracer experiments were identified as being: (i) support for hydrogeological and flow models; (ii) support for descriptions of transport pathways; (iii) support for descriptions of interactions between water, solute and rock mass (including the transfer of laboratory data to the field); (iv) testing/calibration of migration models, particularly in relatively simple systems such as the Boom Clay; and (v) development of overall understanding and confidence building, including public acceptance. These topics were addressed by Working Group 2 (*Rationale and Promises of Future Field Tracer Experiments*).

- **To evaluate the use and results of field tracer tests in the light of alternative testing methodologies.**

The use of tracer-test results for the purpose of performance assessment necessitates transfer, scaling and extrapolation of the results, in both a spatial and temporal sense, even for tests performed within a proposed disposal site. Results of "conventional" field tracer experiments (i.e. relatively short-term tests, using simple tracers, with injection/withdrawal in forced hydrogeological conditions) form an important component of the information required by performance assessment. This information must, however, be complemented by other types of information collected by alternative methods. In this respect, the greater use of natural tracers and natural flow conditions (use of paleohydrogeological and paleohydrogeochemical information) was seen to be crucial. The gap in time scales can partly be filled by long-term experiments which require long-term planning in the programmes. To address the complexity present in many systems, the need for a more sophisticated approach and integration of different approaches was identified. Integration of results obtained using various techniques, including techniques to characterise the site of tracer experiments and identify flowpaths, can give a deeper insight and increased confidence in overall understanding of the migration problem at a given location. This topic was addressed by Working Group 3 (*Alternative Methods to Tracer Experiments*).

- **To assess the results of past and current tests and their uses/relevance for site characterisation/evaluation and performance assessment purposes.**

The workshop noted an increased integration of field tracer experiments and performance assessment in many national programmes. Field tracer experiments can, in certain cases, provide hard information for performance assessment. There is, however, also an increased appreciation of the qualitative aspects of performance assessment to which field tracer experiments can also contribute, providing confirmation that the methodologies, models, processes and data used in performance assessment are appropriate. It was also noted that the contribution of field tracer experiments to performance assessment depends on the stage reached within the waste-management project. This topic was addressed by Working Group 4 (*Integration of Data from Field Tracer Experiments into Performance Assessment*).

- **To compare "generic" tests with site-specific tests.**

Generic tests were considered to be necessary in the study and understanding of processes and transferability of data (e.g. laboratory sorption data to *in situ* retardation). Generic tests are well suited to study the completeness of modelled processes and to identify any omitted phenomena, since the ability of models to reproduce the observed results can often be better tested in generic experiments, where extensive pre-test characterisation and "*post-mortem*" investigations can be performed. Generic tests were seen to complement site-specific tests; site-specific experiments can be interpreted with more confidence if supported by understanding gained in generic tests.

2.2 GENERAL CONCLUSIONS

The following general conclusions have been drawn from the first GEOTRAP workshop:

1. **Field tracer experiments have now been taking place for many years. They have a valuable role to play in building confidence in the identification of the processes relevant to transport, the definition of site models and the provision of parameter values that are required by transport models.**

Tracer tests have demonstrated the operation of matrix diffusion, that current understanding of transport processes is adequate to provide an interpretation of test results and that methods exist that allow laboratory sorption and diffusion data to be applied in the field. For site characterisation purposes, they can support and refine models of particular geologic features and complement conventional hydraulic tests.

2. **The workshop has provided evidence that the radioactive waste community has become more aware of the complexity of the geological environment within which the tests are performed and of the limitations in the applicability of such tests in performance assessment and site characterisation. Where geological complexity is not, however, too great, information can be provided by field tracer experiments that is difficult or impossible to obtain by other means.**

Examples of the kind of data that can be obtained by the modelling of field tracer experiments, but are difficult or impossible to obtain by means such as hydrogeological characterisation (which includes traditional hydrological information, as well as geological characterisation), include flow porosity and, of particular relevance to performance assessment in fractured media, the heterogeneity of flow within fractures (e.g. the degree of channelling). The requirement to test methodologies for transferring laboratory sorption data to field systems can also be met by field tracer experiments. In order to ensure that the conditions of a field tracer tests experiment are as well defined as possible, so that interpretation can focus on a small number of unknowns, the geological setting of the tests should be extensively characterised as part of the planning of the experiments. This task is made easier where geological complexity is not too great.

3. **The presentations and discussions of the workshop have highlighted fundamental differences in field tracer experiments, as applied to plastic clays on the one hand and fractured media on the other, in terms of the information that they provide for performance assessment.**

Field tracer experiments are most successful for simple systems, as exemplified by the Boom Clay tests. The results of field tracer experiments are most easily incorporated into performance assessments, as data for models ("hard" information), when the system is simple in terms of the number of processes operating and in terms of structure (in particular, in the absence of fractures).

Success in the modelling of field tracer experiments in the Boom Clay may have implications beyond geosphere transport modelling in plastic clays. In particular, the techniques developed could be used to test the performance of emplaced bentonite (intact, unjointed blocks) as an effective barrier to the transport of radionuclides. It also opens prospects for the testing of less plastic clays at depth, where fractures may be closed due to the weight of the overburden and may cease to constitute flow pathways.

The link between field tracer experiments and performance assessment is less evident in fractured media. Indeed, such experimental data have found only limited direct use in performance assessment analyses to date. Compared to relatively homogeneous media such as Boom Clay, a larger number of features may be relevant to flow and transport in fractured media and characterisation is frequently incomplete (particularly, characterisation of large-scale heterogeneity). This can result in more complex break-through curves and in the existence of alternative models that fit the curves equally well. It also means that experiments tend to focus on individual features within more complex systems. Nevertheless, modelling exercises can be used to provide support for certain aspects of performance assessment models, such as the averaging of small-scale heterogeneities (see the discussion of "soft" information, below). Furthermore, it is not always necessary, for the purposes of performance assessment, to discriminate between alternative models; for example (i) if a conservative treatment is felt to be acceptable and (ii) if the feature/process concerned is, in any case, insignificant on the spatial and temporal scales of performance assessment.

The usefulness for performance assessment of future tracer tests in fractured media may be increased by carrying out experiments in the structural features that are most relevant to geosphere performance.

4. **There is increased recognition that performance assessment makes use of a combination of quantitative ("hard") and qualitative ("soft") information[1]. Where the system studied is relatively simple (as in the case of plastic clays), field tracer experiments can serve to provide specific hard information. For more complex systems, field tracer experiments can play a useful part in general confidence building, as well as in the development of the team (modellers and experimentalists) and the tools (analytical and experimental) for performance assessment.**

The acquisition of hard information from field tracer experiments is exemplified by the tests carried out in the Boom Clay. The information obtained from the other tests discussed at

---

[1] "Hard" information is, for example, data that can be input, possibly via an interpretative model, into calculational tools. "Soft" information is, for example, wide-ranging evidence that gives confidence that safety assessment methodologies, models, processes, data and system general understanding are appropriate.

GEOTRAP contributes more to general confidence building, particularly in models and the up-scaling of data:

– *Models of processes*: the success of the models used to predict the results of field tracer experiments can build confidence that all processes relevant to solute transport have been identified.

– *Data and up-scaling*: it was pointed out that field tracer experiments are, in general, just one of a number of sources of data, complementing the information from laboratory tests, field hydrogeological tests, natural analogues, natural tracers and geological data. Some of these data sources relate to systems characterised by different scales of space and time to those of interest in performance-assessment calculations (e.g. small-scale laboratory experiments). The application of these data involves assumptions regarding up-scaling. Confidence can be built in these assumptions by demonstrating that they allow the prediction of the results of field tracer experiments, that represent scales intermediate between those of laboratory tests and those of performance assessment, and between those of natural analogues and those of performance assessment.

Field tracer experiments can serve to gather and focus interdisciplinary expert input (for example, through discussion forums such as GEOTRAP). They can also stimulate the development of new experimental techniques (*in situ* and laboratory) and models and establish effective communication between field and laboratory experimentalists and modellers.

5. **The greater complexity and more qualitative link to performance assessment in the case of fractured media suggests that special efforts are required with respect to (i) integration with other types of studies, (ii) the characterisation of the system, (iii) the testing of geological features that are relevant to geosphere performance and (iv) the testing of alternative hypotheses. When testing alternatives, falsified hypotheses should be reported, as well as those that provide successful predictions.**

The approach advocated in the discussions of the GEOTRAP workshop is to reduce uncertainty in the characterisation of the system, as far as is possible, by taking account of, for example, all available geological, hydrogeological and geochemical information and then to test as many alternative models as possible, consistent with the characterisation, against field tracer experiments. Model predictions should be made in advance and tests designed in such a way that they can differentiate between alternative models, thus allowing hypotheses to be falsified. Even if a thorough hydraulic characterisation is a prerequisite for a good experimental design and a proper interpretation of the test, it is advisable, where possible, to analyse the flow system in detail (over-coring or moulding of the fracture system) after the completion of the tracer experiments. This may help fix some of the free parameters that the models may contain.

6. **The similarity, in terms of processes, of the models that are applied in the interpretation of field tracer experiments for a particular medium suggests that a consensus may have been reached in the identification of processes relevant to tracer transport.**

In practice, it is rare in the modelling of field tracer experiments to examine conceptual model uncertainty in the sense of uncertainty in the processes that are operating. Rather, alternative models tend to represent alternative ways of simplifying a geological interpretation of the system in order that transport modelling may be performed. Recent experience in field tracer experiments has suggested that no additional processes, outside the scope of current models, need to be invoked in order to understand experimental results.

7. **Efforts are required in the communication (i) of performance assessment requirements to experimentalists involved in field tracer experiments, (ii) (by modellers) of the need for simplification of geological representations and (by experimentalists) of the extent to which such simplifications are geologically meaningful, and (iii) of key results to programme managers, regulators and the public.**

A number of suggestions were made at the GEOTRAP workshop as to how to achieve such communication. Among these were:

– Overlap, where possible, of the teams involved in field tracer experiments and the teams involved in performance assessment. This overlap was advocated both on the modelling side and on the experimental side.

– Information exchange in a form that is as simple and concise as possible, consistent with its intended audience and application.

In principle, it was agreed that performance-assessment teams should be involved in the identification of experiments to be performed and in the design of these experiments, particularly in the later stages of a repository research programme. The generic experiments performed in the early stages of a repository research programme have provided input to performance assessment, providing support for the basic understanding of transport processes. In the later stages of repository development, where the emphasis shifts to the refinement and testing of models of transport at a specific site, the input of performance-assessment teams to the planning of experiments will become increasingly important. This will ensure that the experiments, in conjunction with detailed hydrogeological characterisation, deliver information (flow porosity, channelling, etc.) that is required in order to assess the performance of the geological barrier.

2.3   LIMITATIONS OF TRACER TESTS

The limitations regarding the use of field tracer experiments in support of the assessment of the geosphere as a barrier to potential radionuclide releases from a deep repository were clearly acknowledged during the discussions. These limitations may be encountered at the licensing process level, or when planning the strategy by which to provide evidence of the fulfilment of safety criteria for waste disposal. They may also be encountered at the level of technical details of experimental procedures. Much progress has been made recently, especially in the use of tracer tests to characterise heterogeneous media and to study transport phenomena therein. Some limitations that are inherently associated with field tracer experiments or that remain regardless of the latest developments will be discussed here.

• **Field tracer experiments (like any other test) can provide information only on the present (and, to some extent, past) situation; long-term changes and their effects have to be assessed by other means.**

The geosphere has gone through many evolutionary changes and cycles and will continue to do so in the future. The relevance of present-day determination of structures and properties of transport pathways for a case in the distant future has to be assessed critically. On the other hand, in many cases, conditions deep in geologic formations can be shown to have been stable over long periods of time.

- **The extent to which a repository site can be characterised by means of field tracer experiments depends on the characteristics, location and scale of the planned tests.**

The thorough study of a site associated with large-scale heterogeneities is, in practice, impossible. The number of boreholes needed for such a characterisation would be so large that, even if hypothetically possible from an economic perspective, the existence of the boreholes would destroy the natural conditions at the site. The characterisation is inherently limited to certain specific areas. Nevertheless, this may prove very valuable.

- **The time and length scales of field tracer experiments compared to those relevant in assessing the performance of the geosphere is problematic. Only a limited volume of a host formation and its surrounding can be covered by rather short-term tracer tests.**

Tracer tests have mostly to be performed in forced flow conditions over scales such that reasonably high recoveries can be achieved in rather limited times. A combination of long transport paths and slow flows in an unknown flow field is not possible. For sorbing tracers, the test times would exceed any practicable limits. Particular components of the potential migration paths can, however, be examined and provide a basis for the assessment of complete paths. The length scales that have been used experimentally are, however, very relevant for an analysis of the transport properties of the near-field host rock.

- **The extent to which chemical and physical disturbances caused by the tests themselves can be avoided needs to be carefully assessed.**

The forced flows that are unavoidably used in the experiments, as well as equipment in the test area, cause a disturbance in the test environment that may be difficult to estimate. Development work aimed at minimising disturbances has been, and will continue to be, performed. There is, however, a trade-off between low disturbance and the desire for experiments covering large spatial scales.

- **Non-uniqueness of interpretation.**

Breakthrough curves from tracer experiments are often difficult to interpret in relation to the different processes operating along the flowpath, and a unique interpretation (say, in terms of matrix diffusion) is rarely possible.

- **The optimum way in which to integrate field tracer experiments with other studies and with performance assessment may be unclear.**

Tracer experiments are only one source of information relevant to performance assessment. To get the best out of tracer tests, that are often long-lasting and expensive, *they have to be integrated with other characterisation work* and should serve the needs of the performance assessment. It has proved difficult, in some instances, to find a common forum and language for discussions between experimentalists and performance assessors. It is not an easy task to simplify a complex flow and transport path structure, as well as the sometimes complex phenomena taking place within it, as a clear and not excessively conservative conceptualisation. Advances towards such conceptualisations are best achieved in a dialogue between experimenters and performance assessors.

- **The circumstances when, and reasons why, one should perform field tracer experiments must be identified.**

A tracer test is by no means an overall remedy to every problem, but is an effective tool to study well-defined problems of flow and transport. It can be most useful when supported by as much information as possible, from other techniques, about the system under study. This supporting information may prove to be sufficient in itself, circumventing the need for a tracer test. In some cases, however, there may be specific open questions to which tracer tests are well suited. The rationale will, in such cases, be well formulated and the test design can be based on focused objectives. In other cases, the usefulness of tracer tests should be considered carefully. The role of tracer tests in different programmes and in their different phases varies considerably. It should always be an open question whether a tracer test is a suitable tool in any given circumstances.

2.4   RECOMMENDATIONS

2.4.1   Usefulness of Field Tracer Experiments

Field tracer experiments should continue to form part of the research needed for performance-assessment modelling as they provide important technical information, help build confidence in performance assessment models, and provide much, sometimes unexpected, supporting information and understanding of radionuclide-migration mechanisms. Among the wider benefits to be derived from the continued use of field tracer experiments are also the development of interdisciplinary teams and improvement of public confidence in disposal options. Though it should be remembered that the results of field tracer experiments can be difficult to interpret and that such tests cannot provide proof of the performance of the geological barrier of any disposal system, but are rather a technique among others to be used in pursuit of a solution to the safe disposal of radioactive waste.

- **Technical input to performance assessment models:** Field tracer experiments can serve at several levels in safety-related studies of geologic disposal systems for radioactive waste. Tracer tests sometimes provide the only reference to solute migration under real field conditions and can be used for various purposes at different scales. A good tracer test provides the desired information and data, and confirms (or, perhaps, denies) the anticipated and modelled behaviour of solutes and radionuclides in geologic media.

- **Confidence building:** Field tracer experiments play an important role in building confidence in performance-assessment models and methods and enable realistic models to be developed (or, at least, contribute towards their development). Field tracer experiments are most useful when integrated with other investigations as part of an overall programme; they thus complement other techniques and should not be seen as an alternative to any other type of investigation.

- **Supporting information:** Field tracer experiments can provide a wide range of both hard (i.e. quantitative) and soft (i.e. qualitative) data. The quantitative results of tracer tests can be fed directly into model simulations of tracer behaviour. The importance of qualitative information should not be underestimated, however, as this kind of "understanding" places constraints on, and confidence in, the transport-model development. Unexpected results of tracer tests can be particularly instructive, especially when they reveal unrealistic assumptions.

- **Development of interdisciplinary teams:** Tracer experiments provide a focus for a wide range of studies and act as a spur for the establishment of interdisciplinary teams, including field scientists (geologists and hydrogeologists), experimentalists, modellers and performance-assessment specialists. From this should arise a more complete conceptualisation and understanding that is necessary if modelling of the complexity of radionuclide migration in geologic media is to be tackled successfully.

- **Public confidence:** Field tracer experiments not only build confidence in the models, but can also contribute to public confidence that the level of understanding of radionuclide transport necessary for radioactive waste management is being attained; this, in turn, contributes to the public acceptance of waste management programmes in general.

2.4.2   Design and Performance of Field Tracer Experiments

In designing tracer tests, and in interpreting test results, special attention should be paid to the degree to which the data obtained are representative of the system of interest and to disturbance to the system by the test itself, while full advantage should be taken of the experience from earlier tests:

- **Degree to which data are representative:** Parameter uncertainty should be evaluated considering:

  - that the values of any parameters inferred from the results of the test are specific to the scale of the test and may represent some average over the region of the test (e.g. concentration values will be averages over a sampling interval and some volume around the interval);
  - that the values will be specific to the domain in which the test was performed. This means that, if the geological medium exhibits significant variability on a larger scale, the test could yield very different results for a similar domain at a different location and the test will not enable the larger-scale variability to be characterised. If, however, the results of the tracer test can be related to parameters whose distribution on the larger scale has been characterised, it may be possible to infer the likely results of performing the tracer test at an alternative location.

- **Disturbance to the natural system:** Disturbance to the system caused by the test itself should be considered as this might invalidate the data obtained (the results being related to an artificial rather than natural system) or, at least, make it impossible to reproduce the test using the same flow path.

- **Earlier experience:** Many tracer tests have been performed at many sites during the last two decades, not all of which have been reported at this workshop. There is now some repetition of the work carried out in the past, due to a lack of awareness of previous studies, and the experience gained from these earlier investigations is being lost. In developing new tracer tests, advantage should be taken of the lessons learned from these earlier tests.

- **Relevant radionuclides:** Tracer tests with performance-assessment relevant radionuclides are very rare. As such, any tests using these radionuclides are to be encouraged in order to study their *in situ* behaviour.

ANNEX 1

SYNOPTIC TABLES OF THE TESTS DISCUSSED

Several field tracer experiments were described in the course of the presentations and discussions. In order to set the context in which the achievements, general conclusions, open issues and recommendations should be viewed, Tables 1 and 2 summarise the main characteristics of these experiments, including their current status and principal aims.

Table 1. **Geographical locations, geological media and status of field tracer experiments discussed at the GEOTRAP workshop**

Test/ organisation	GEOTRAP reference(s)	Geographical location	Geological medium	Status
Tracer tests in **Boom Clay** / SCK-CEN.	Session II: VOLCKAERT & GAUTSCHI	HADES underground research facility, near Mol, NE Belgium.	Plastic, Tertiary clay (Boom Clay).	Diffusion tests with tritiated-water and ^{125}I tracers completed. 3-D tests using ^{14}C-labelled bicarbonate started in 1995. Further tests planned with ^{14}C-labelled organic molecules.
International **Mt. Terri** Project / Andra, BGR, Enresa, Nagra, Obayashi, PNC, SCK-CEN, SNHGS	Session II: VOLCKAERT & GAUTSCHI	Mt. Terri motorway tunnel near St. Ursanne in the Jura mountains, NW Switzerland.	Minor faults and fractures in a well-consolidated Middle Jurassic shale (Opalinus Clay) in the southern limb of Mt. Terri in the Folded Jura tectonic unit.	Feasibility study in progress (laboratory experiments, improvements of over-coring technique).
H-19 and H-11 Tracer Tests at the **WIPP** site / SNL, DOE	Session II: BEAUHEIM et al.; Session III: MEIGS et al.	Waste Isolation Pilot Plant (WIPP) site in the Delaware Basin, SE New Mexico, USA.	Fractured Permian Culebra Dolomite (Rustler Formation).	Field single-well injection-withdrawal and multiwell convergent-flow tests with non-sorbing tracers completed; planning underway for laboratory diffusion tests.
Grimsel Migration Experiment / Nagra, PNC.	Session II: ALEXANDER et al.; Poster Session: ALEXANDER et al.	Grimsel Test Site, east flank of the Juchlistock mountain in the Aar Massif of the Central Swiss Alps.	Reactivated mylonitic shear zone in Carboniferous Grimsel Granodiorite.	Numerous tracer tests with weakly and moderately sorbing tracers completed. Tests underway with more strongly sorbing tracers and with subsequent "*post mortem*" excavation of a portion of the shear zone.
Tracer tests at the **URL** / AECL	Session III: FROST et al.; Session IV: FROST et al.	AECL's Underground Research Laboratory, Lac du Bonnet, Manitoba, Canada.	Fractured crystalline rock, including fracture zones, moderately fractured rock, sparsely fractured rock and excavation damaged zones.	Two-well tracer tests completed within several major low-dipping fracture zones. Tracer tests underway in a region of moderately fractured rock with interconnected networks of discrete fractures. Migration experiment (in co-operation with JAERI) also underway in natural fractures in excavated granite blocks. Tracer experiment has been conducted within excavated damaged zone of a test tunnel.

Table 1. (continued from previous page)

Test/ organisation	GEOTRAP reference(s)	Geographical location	Geological medium	Status
Äspö HRL TRUE Programme/ SKB	Session III: OLSSON & WINBERG	SKB Äspö Hard Rock Laboratory, SE coast of Sweden.	TRUE-1: reactivated mylonitic shear zone in the Äspö diorite. TRUE Block Scale: Fracture network in a rock consisting mainly of Äspö diorite.	TRUE 1: radially converging and dipole expts. using conservative tracers completed. To be followed in 1997 with expts. with sorbing tracers, followed by resin injection and excavation. TRUE Block Scale: suitable site selected, pilot borehole drilled and characterised, preliminary site characterisation in 1997.
Tracer tests at the **El Berrocal** Site/ Enresa, EU.	Session III: GUIMERÀ et al.; Poster Session: GARCIA-GUTIÉRREZ et al.	El Berrocal Site, Central System, Central Spain.	Fractured granite.	7 tests carried out, with tracer recovery from 4 of them. Project complete. No more tests planned.
Tracer experiment at the **Kamaishi Mine** / PNC	Session IV: UCHIDA et al.	Kamaishi Mine in the Kitakami Mountains, Iwate Prefecture, Northern Honshu, Japan.	Cretaceous granodiorite.	Final borehole array completed and site-characterisation on-going to determine candidate fractures. Tracer-test design in progress; tracer tests to be conducted May 1997 to March 1998.
Combined pumping and tracer test at **Palmottu** / GTK, EU	Session IV: GUSTAFSSON et al.	Palmottu study area, SW Finland, within a zone of metamorphosed schists and gneisses that extends from southern to central Finland.	Fracture zones in crystalline rocks (mica gneiss and granite), surrounding uranium mineralisation.	Test is in the planning stage; detailed design is under discussion.
Radionuclide migration following nuclear explosions in **rock salt**/Radium Institute	Poster Session: ANDERSON et al.	Great Azgir salt dome, SW Caspian Sea depression, Kazakhstan.	5 stable cavities filled with radioactive brine. Permian rock salt diapir.	60s - 70s: comprehensive surveys during the conduction of the nuclear explosions. 80s - 90s: radiochemical monitoring. Current: feasibility study: analogue for studying the isolation capacity of rock salt.
Tracer tests at the **Reskajeage Quarry**/AEA, Nirex	Poster Session: HOLTON et al.	Cornwall, UK.	Fractured slate.	Combined colloid and non-sorbing tracer migration tests. Complete.

Table 2. Scales, tracers used and principal aims of field tracer experiments discussed at the GEOTRAP workshop

Test/ organisa-tion	Scale of tests		Tracers		Principal aims
	temporal	spatial	conservative*	sorbing	
Tracer tests in **Boom Clay**/ SCK-CEN.	7 years - still on-going.	A few metres.	Tritiated water, ^{125}I, ^{14}C-labelled bicarbonate, ^{14}C-labelled organic molecules.	None used.	(i) demonstration of the predictability of radionuclide migration in the Boom Clay and assessment of the reliability of these predictions; (ii) enhancement of public acceptance.
International **Mt. Terri** Project / Andra, BGR, Enresa, Nagra, Obayashi, PNC, SCK-CEN, SNHGS	Tracer injection in a packed-off section of a single, small-diameter borehole over a long period (2 years or more).	Either over-coring of injection borehole or drilling of a parallel borehole at a distance of one to a few metres.	Not yet fixed (on-going feasibility laboratory experiments at CEN/DAMRI, Grenoble, France and at the University of Berne, Switzerland).		(i) visualisation of flowpaths; (ii) identification of groundwater flow and solute transport mechanisms in a highly-consolidated, fractured claystone; (iii) evaluation of parameters for radionuclide transport models.
H-19 and H-11 Tracer Tests at the **WIPP** site/SNL, DOE	Single-well injection-withdrawal tests: 18-hr pause after injection, 20-40 days pumping; convergent-flow tests: 14-105 days.	Convergent flow fields with travel paths of 11 to 25 m.	Iodide, 4 dichlorobenzoic acids, trichlorobenzoic acid, 6 difluorobenzoic acids, 2 trifluorobenzoic acids, tetrafluorobenzoic acid, pentafluorobenzoic acid, 3 trifluoromethyl-benzoic acids.	None used	(i) test for the occurrence of matrix diffusion in the Culebra; (ii) quantify or bound the amount of matrix diffusion occurring; (iii) evaluate whether or not idealised uniform fracture-matrix geometry is adequate to model test results; (iv) evaluate the effects of layering within the Culebra on flow and transport; (v) investigate the causes of directional differences in transport within the Culebra.
Grimsel Migration Experiment / Nagra, PNC	Min ~ 1 week; max ~ 20 months. Excavation project: injection & dipole pumping ~ 4 weeks.	Dipole flow fields with distances from injection to extraction of 1.7, 4, 5 and 17 m.	Uranine, ^{3}H, $^{3,4}He$, $^{82}Br^-$, ^{123}I.	$^{22,24}Na^+$, $^{85}Sr^{2+}$, $^{86}Rb^+$, $^{137}Cs^+$, $^{99m}TcO_4^{2-}$. Excavation project: ^{99}Tc, ^{79}Se, ^{152}Eu, ^{237}Np, ^{113}Sn, Mo (stable), ^{60}Co (^{63}Ni), $^{234,235}U$.	(i) study of the hydrology and geochemistry of a fractured rock; (ii) testing of models of radionuclide transport; (iii) development of methodologies for site characterisation; (iv) focusing of laboratory, field and modelling studies to the detailed characterisation of a single site.

*It is noted that all tracers, even those classed as "conservative", display some interaction with the geological medium through which they migrate.

Table 2. (continued from previous page)

Test/ organisa-tion	Scale of tests		Tracers		Principal aims
	temporal	spatial	conservative	sorbing	
Tracer tests at the **URL** / AECL: Fracture zones	~ 1 week to 8 months.	~ 20 m to 700 m	I^- and Br^-; colloid tracers.	None used.	Determination of the physical solute transport properties of volumes of intensely fractured rock and development of methods for extrapolating the test results to larger scales.
Tracer tests at the **URL** / AECL: Moderately fractured rocks	~ 1 week to several months.	~ 10 m to 50 m in a large volume (~ 10^5 m^3) of rock	I^- and possibly others; colloid tracers.	Currently under study.	(i) evaluation of the physical and chemical solute transport properties of a relatively large volume of moderately fractured rock; (ii) determination of the suitability of the porous-media-equivalent method for modelling solute transport in volumes of moderately fractured rock; (iii) evaluation of other modelling approaches such as discrete fracture network models.
Tracer tests at the **URL** / AECL: Sparsely fractured rocks	~ 1 week to ~ 1 year.	~ 1 m	^{3}H.	^{85}Sr, ^{95m}Tc, ^{237}Np, ^{238}Pu.	Study of the transport of conservative and sorbing radionuclides in natural fractures in 1 m^3 quarried blocks of granite under *in situ* groundwater conditions.
Tracer tests at the **URL** / AECL: Excavation damaged zones	~ 2 days.	1.5 m	I^-.	None used.	Acquisition of information on physical solute transport properties within excavation damaged zones surrounding underground tunnels.
Äspö HRL TRUE Programme / SKB.	Hours - months.	Lab: <1 m; detailed: < 10 m; block scale: 10 - 100 m.	Uranine, Eosine, Rhodamine, Amino-G, metal complexes.	Selection of radioactive cations among: Na, Ca, Sr, Rb, Ba, Cs.	(i) development of understanding of radionuclide migration and retention in fractured rock; (ii) evaluation of the extent to which concepts used in models are based on realistic descriptions of rock and of whether adequate data can be collected in site characterisation; (iii) evaluation of the usefulness and feasibility of different approaches to radionuclide migration and retention; (iv) provision of *in situ* data on radionuclide migration and retention.
Tracer tests at the **El Berrocal** Site / Enresa, EU	Min ~ 1 week; max ~ 3 months.	8 m to 25 m.	Uranine, Eosine, Brillant Sulphaphlavine, Iodide, ^{82}Br$^-$, Gadolinium, Rhenium, Phloxine, ^{2}H.	None used	Hydrodynamic characterisation of main geological structures.
Tracer experiment at the **Kamaishi Mine** / PNC	To be determined on the basis of on-going scoping calculations.	2 m to 60 m.	Uranine + others not yet fixed.		(i) to obtain a conceptual model with realistic geometries and properties of conductive fractures; (ii) to understand the hydraulic properties and geometries relevant to flow; (iii) to test a discrete-fracture model; (iv) to develop a site-characterisation methodology to be used for an eventual Japanese deep repository.
Combined pumping and tracer test at **Palmottu** / GTK, EU	2 - 4 weeks.	Converging flow field, with distance from injection to extraction of 20 - 100 m.	Fluorescent dyes (uranine, amino-G, rhodamine WT) and stable metal complexes (Gd-DTPA, Ho-DTPA, Eu-DTPA).	None will be used.	(i) verification of the updated conceptual hydrostructural model around the central part of the U-mineralisation; (ii) identification of the main potential flow paths within the test domain; (iii) improvement of understanding of flow and transport properties in order to support the forthcoming analogue transport study.

Table 2. (continued from previous page)

Test/ organisa-tion	Scale of tests		Tracers		Principal aims
	temporal	spatial	conservative	sorbing	
Radionuclide migration following nuclear explosions in **rock salt**/Radium Institute	Nuclear explosions detonated between 1966 and 1979. Annual sampling of brine 1980 - 1991.	Sampling of brine at 13 - 200 m (horiz.) from explosion epicentre.	Numerous fission and activation products and residual fissile isotopes of Uranium and Plutonium. Migration in brine of ^{137}Cs and ^{90}Sr studied in detail.		(i) assessment of rock salt as a medium for isolation and disposal of radioactive waste; (ii) evaluation of underground nuclear explosions as large-scale geo-technical analogues of radioactive waste disposal.
Tracer tests at the **Reskageage Quarry**/AEA, Nirex	?	5, 9.4 and 15.4m	Colloids: mono-dispersed silica particles and monodispersed hematite. Dye: rhodamine - wt.	None used.	(i) evaluation of the mobility of different types of colloids in a fractured rock environment; (ii) confidence building in modelling transport processes.

ANNEX 2

DETAILED RECORD OF THE WORKSHOP

1.   INTRODUCTION

The first GEOTRAP workshop took the form of a series of oral and poster presentations, followed by discussions by four specific working groups and a final, general discussion session. This chapter compiles the answers, provided in each presentation, to the key questions specified for the four workshop sessions, and synthesises the discussions and conclusions of the four working groups and the final, general discussions.

The oral presentations were organised into four sessions:

- Session I: *General Overview;*
- Session II: *Rationale Behind Field Tracer Experiments;*
- Session III: *Test Cases: Design, Modelling and Interpretation;*
- Session IV: *Aims and Design of Planned Field Tracer Experiments.*

For each overview paper in Session I, in order to maintain focus, specific questions were set, in advance of the workshop, for the authors to address. In Sessions II-IV, questions were set for each session, and all the papers presented within a particular session aimed to address those questions. Section 2, below, indicates how the various papers addressed these questions. The answers provided aim at presenting the authors of the papers points of view and do not especially constitute a consensus statement that was reached at the end of the workshop.

More technical details of field tracer experiments were presented as posters. The presented posters are not included in this record of the workshop.

The four working groups focused on the following key aspects of field tracer experiments:

- Working Group 1: *Practical Challenges;*
- Working Group 2: *Rationale and Promises of Future Field Tracer Experiments;*
- Working Group 3: *Alternative Methods to Tracer Experiments;*
- Working Group 4: *Integration of Data From Field Tracer Experiments into Performance Assessment.*

A series of key questions was also specified for each working group in advance of the workshop. Section 3 presents the conclusions drawn by each of the working groups. These conclusions do not especially represent consensus statements that were reached at the end of the workshop.

Finally, Section 4 provides a summary of issues that were addressed in the final discussion session of the workshop.

2. RECORD OF THE WORKSHOP SESSIONS

2.1 Session I: General Overview

The GEOTRAP Programme Committee set specific questions to be examined in the papers and presentations. It should be remembered, that it is impossible to cover all tracer tests with one general answer. The answers should be seen as representing a trend among the tests that have been performed, rather than being applicable to any individual experiment.

What Has Been Learned from Field Tracer Transport Experiments – A Critical Overview

The paper by HAUTOJÄRVI, ANDERSSON & VOLCKAERT addressed the following questions.

1. Was the rationale of these tests clear enough?

The rationale for field tracer experiments has generally been clear, but the aims are often too optimistic and unrealistically wide, when account is taken of the available resources. For porous media, the typical rationale is the need to know the flow and transport porosity, together with the quantification of dispersion in one, two, or three dimensions. It is a simple and clearly stated rationale. Required test arrangements and procedures may, however, be quite complicated. The rationale for experiments in fractured media have often been expressed in similar terms. This may be a severe problem if the analogy between the media does not hold. A characteristic phenomenon of fracture flow is channelling, and the rationale of many experiments has been based on this point. Experimentally, it is a challenge to address channelling in undisturbed rock and near-natural flow conditions.

2. Which information was really obtained? What have field tracer tests taught us about important transport mechanisms?

Flow velocities and dispersion can usually be obtained quite reliably and accurately. Beyond that, the information obtained depends much on the concepts and modelling used. Usually, there are many different concepts and models that can, at the same time, explain a given set of results. There are thus ambiguities in the interpretation, which cannot be always be resolved due to lack of experimental data. The governing transport mechanisms cannot be distinguished in such cases. The most debated, and perhaps also most important transport mechanism is the matrix diffusion. It is extremely difficult, if not impossible, to show the effect of matrix diffusion on the break-through curves of field tracer experiments. It is certainly not enough to fit an advection-dispersion-matrix diffusion model to a break-through curve and deduce the various transport mechanisms from the model parameters.

3. What use was made of the information obtained?

The information has mostly served to strengthen understanding of the flow in different media: transport calculations in performance assessments are based directly or indirectly on the flow characteristics. Still, there are uncertainties in the basis of transport modelling, as partly discussed under question (2).

4. Have they helped to build confidence in the predictive modelling of radionuclide transport for performance assessment purposes?

Tracer tests have partly helped to build confidence in transport modelling. There are, however, important gaps to be filled before a satisfactory level of confidence can be achieved.

5. Where are the failures? Were these failures clearly reported? What are the lessons learned from them?

Where tests have failed to fulfil their goals regarding understanding of transport mechanisms, this has mainly been due to practical test limitations and partly also to unfavourable test procedures for the given goals. The reports usually emphasise a good agreement of the experimental and modelled results, and possible ambiguities are not assessed critically. In this sense, the "failures" or insufficiencies of the tests are hidden, rather than clearly reported. The lessons learned from failures of these kinds are that more and better tests and detailed characterisation of the test site is needed before transport mechanisms can be revealed and studied in the field.

6. What can be expected from future field tracer tests?

In future tests, ambiguities are likely to be reduced and tests are likely to be more focused on specific transport mechanism studies, compared to the "overall" type of tests made in the past. This may mean the performing of various tests at various scales and the combining of results from different tests (site, generic and laboratory). The aim should be to distinguish different concepts and models by the comparison of predictions with test results. It seems that not all of the tests that might be wished for can be performed in the field at a specific site. The tests at a disposal site will be even more limited in number. Characterisation of hydraulics is the main task at a disposal site and performance assessment has to rely on the relations between the hydraulic and transport properties studied at other places, possibly nearby, and even on generic studies.

The Contribution of Field Tracer Transport Experiments to Repository Performance Assessment

The paper by SMITH & ZUIDEMA addressed the following questions:

1. What are the PA needs for field tracer tests? What kind of answers can field tracer tests provide?

PA needs are identified as:

- Confidence building and identification of uncertainties: the success of a model in reproducing the results of field tracer tests, well-designed for specific purposes, builds confidence that relevant features and processes have been identified. Confidence is also built in the methodology for quantifying the rates of processes and the spatial extents over which they operate. The application of alternative models is useful in indicating the degree of conceptual uncertainty and, where some models fail, in narrowing the range of uncertainties.

- Assessment model formulation: assessment models are frequently derived from the more complex models used to interpret field tracer tests. Understanding of tests by means of models that aim at realism can ensure that the simplified assessment models represent key structures and processes, that simplifications do not give non-conservative results and that laboratory and field data are used appropriately.

In addition to the above roles, it is pointed out that, in a few cases, tracer tests have been used directly in the characterisation of flow and transport properties at specific sites.

2. How can they help to build confidence in the predictive modelling of radionuclide transport for predictive purposes?

A distinction is drawn between "inverse modelling" of tracer tests, which is used for model calibration, and predictive modelling of tracer tests, which is generally more convincing in terms of confidence building in models that will be used (perhaps in simplified form) to predict radionuclide transport in PA

In order to maximise benefits from predictive modelling of tests, it is recommended:

- that predictions be made *in advance of the tests*, with clear concepts, with a methodology defined for setting parameter values and with "success criteria" that take account of experimental errors; and

- that as wide a range of plausible alternative models as possible are examined.

3. How are the results of these tests used in performance assessment?

The primary uses for field tracer experiments in PA, the needs of which are indicated in (1) above, are:

- confidence building and the formulation of models, in which tracer tests have illustrated the importance of an understanding of small-scale geological structure and demonstrated the operation of matrix diffusion. They have also demonstrated that current understanding is adequate to provide an interpretation of many test results and that methods exist that allow laboratory sorption and diffusion data to be applied in the field.

- site characterisation, in which the use of tracer tests is restricted due to large-scale heterogeneity (see 4, below) and possible perturbation of the site by the tests themselves. Tracer tests can, however, be used to support and refine models of particular geological features and the related "tracer-dilution tests" can complement conventional hydraulic tests.

Additional benefits from tracer tests include experience in the practicalities of obtaining field data and relevant laboratory data, the development and refinement of measuring devices and the establishment of successful communications between geologists, laboratory and field experimentalists and modellers.

4. How can results of field tracer tests be extrapolated and/or transferred to larger volumes of rock and to other sites/geology?

- It is acknowledged that tracer tests provide no information on features and processes that, though irrelevant on the spatial and temporal scales of the tests, may be important over scales relevant to performance assessment - i.e. slow processes and processes operating on large-scale features.

- It is suggested that the identification of slow processes is the domain, for example, of natural analogue studies, rather than of field tracer tests.

- In the case of large-scale features, attempts that have been made to model tracer tests in networks of fractures have so far met with only limited success due to the lack of detailed characterisation of the networks.

- It is pointed out that the possibility of "fast channels" through the host rock is a key issue in geosphere PA and that, if their existence cannot be excluded, then the results of tracer tests are of rather limited use.

Regulator's Point of View on the Use and Relevance of Field Tracer Transport Experiments

The paper by BOGORINSKI and BALTES addressed the following questions:

1. Are the tests expected to be helpful in the perceived safety of a repository?

Tracer tests are potentially helpful in the assessment of repository safety in that they can:

(i)   test the "generic" capabilities of groundwater flow and radionuclide transport models (see 2, below);

(ii)   provide certain site-specific data, such as porosities, diffusivities and retardation properties, that are relevant to the modelling of radionuclide migration through the geosphere;

(iii)   provide input to the characterisation of the immediate vicinity of a repository and, in particular, the excavation disturbed zone.

2. How can the tests help to build confidence in the predictive modelling of radionuclide transport?

Confidence in the predictive modelling of radionuclide transport can be enhanced by performing tracer experiments either at an underground laboratory or within geological formations similar to those at a selected disposal site. The latter provides a test of the generic capabilities of such models. Testing in an underground laboratory at an actual site presents practical difficulties (e.g. disturbance of hydrogeological and geochemical conditions). Nevertheless, the site-specific data that such testing provides can be useful for the modelling of particular, spatially-limited features (e.g. the excavation disturbed zone).

3. How useful were previous tests? What was missing?

A problem specific to soft sedimentary rocks, such as those that are relevant to the German waste management programme, is that transport is slow and is not confined to just a few distinctive pathways (as is frequently the case with hard rocks). Field tracer experiments are not viewed by the authors as an appropriate means to characterise the large-scale heterogeneity of such rocks, because recovery of tracers at distant monitoring boreholes would be poor and difficult to interpret.

A more general problem is that, in order to fully understand the migration of a tracer, a network of boreholes is required that inevitably perturbs the system, introducing an artificial heterogeneity. If it were possible to develop novel methods to measure tracer concentrations within a rock without disturbing it with boreholes, this would greatly enhance the usefulness of such tests.

4. What is the relevance of these tests for site characterisation and site evaluation purposes?

It is pointed out that the spatial and temporal scales of relevance to performance assessment may be orders of magnitude greater than those that can be studied in tracer tests and that, in site characterisation, tracer tests can only cover a small part of the region of interest. Tracer tests can, however, be useful in determining hydrogeological and transport parameters at specific locations, in particular where variations due to geological events are suspected and in the excavation disturbed zone.

The authors conclude that they, as regulators, would not request specific tracer tests for the sole purpose of characterising the hydrogeological and geochemical conditions at the site of a proposed nuclear waste repository at a scale relevant to performance assessment.

2.2   Session II: Rationale Behind Field Tracer Experiments

The papers in Session II consider the rationale behind (i) tracer experiments in the Boom and Opalinus Clays (VOLCKAERT & GAUTSCHI), (ii) tracer experiments at the WIPP site (BEAUHEIM et al.) and (iii) the Grimsel Migration Experiment (ALEXANDER et al.). In addition, a paper by VIRA considers the relevance of tracer experiments to site description and understanding and how the benefits of tracer tests can be judged. The following tables indicate how the questions set for Session II are addressed by the papers.

1. What was the general context of the tests and how did this context influence the tests?

Tests	Context	Influence on tests (see also 2, below)
Boom Clay	Safety studies shown that the Boom Clay layer is the most important barrier in the Belgian multi-barrier concept and that diffusion is the dominant transport process.	Emphasis on extrapolation of (diffusion) data from lab to field. Slow transport rates means that use of sorbing tracers is not practical.
Opalinus Clay	Opalinus clay is a potential host rock for a Swiss HLW repository. No formal PA study yet performed.	Emphasis on identification of basic, transport-relevant phenomena (e.g. importance of joints and faults).
WIPP	Review of the 1992 PA for the WIPP site concluded that there was inadequate experimental justification to rule out alternative models and parameters for transport in the Culebra Dolomite Member, which overlies the salt host rock.	Emphasis on provision of a defensible model and parameters for PA modelling of the Culebra Dolomite member.
Grimsel	Fractured crystalline rock is a potential host for Swiss and Japanese HLW repositories. There is a desire to improve confidence in the use of transport codes for such a medium in PA.	Emphasis on testing (in advance of expts.) the predictive capabilities of models.

2.   What were the objectives of the tests and how were these objectives incorporated in the test design?

Objective	Boom clay	Opalinus clay	WIPP	Grimsel
Understanding basic phenomena governing mobility of radionuclides.	×	yes	yes	yes
Direct determination of radionuclide migration parameters.	×	yes	yes	×
Studying transferability of lab data to *in situ* conditions (1).	yes	×	×	yes
Development and refinement of radionuclide transport models for PA.	×	×	yes	yes
Demonstration of predictability of radionuclide migration through a potential host rock.	yes	×	yes	(2)
Enhancement of public acceptance.	yes	×	yes	yes

Notes: (1) The lab experiments in question are chiefly, in the case of Boom Clay, diffusion experiments on small-scale samples and, in the case of the Grimsel Migration Experiment, batch sorption experiments.

(2) Grimsel granodiorite is not a potential repository host rock. Rather, it is regarded as a "generic" crystalline rock. A related objective of the Grimsel Migration Experiment is, however, given as - "... indoctrination of staff into the mind-set required for them to make predictions of radionuclide behaviour *in situ* ... ".

The test designs that have been established in order to address these objectives are described in the individual papers.

3. How does the test relate to the overall R&D programme?

Tests	Relation to overall R&D programme	
Boom Clay	Direct input to laboratory diffusion programme (relationship between lab. data and *in situ* sorption).	
Opalinus Clay (1)	(i)	Feasibility of experiments currently under assessment in laboratory studies.
	(ii)	Outcome of tests will provide input in the formulation of an appropriate transport model for PA
WIPP	(i)	Linked to field hydraulic-testing programme.
	(ii)	Linked to lab. programme (solubilities, batch-sorption studies with crushed Culebra matrix, matrix porosity, tortuosity and permeability).
Grimsel	(i)	Direct input to Nagra's laboratory sorption programme (relationship between lab. data and *in situ* sorption).
	(ii)	"Cross-fertilisation" between experiment and site characterisation/ performance assessment in the field of flowpath description.
	(iii)	Public relations: articles in Nagra Bulletin, production of videos in Switzerland and Japan, public access to Grimsel Test Site.

Note: (1) The Opalinus-clay tests are still at the planning stage.

4. What use was made of the information obtained and how were the results extrapolated?

Tests (1)	Use made of results in *PA*	
Boom Clay	(i)	General confidence building in diffusion-dominated transport in Boom Clay (as assumed in PA).
	(ii)	Conceptual model and data used for the prediction of tracer tests used directly in performance assessment studies (e.g. EVEREST, EC Study on the Evaluation of Elements Responsible for the Effective Engaged Dose Rates Associated with the Final Storage of Radioactive Waste).
WIPP	(i)	General confidence building in the dual-porosity concept for geosphere transport modelling.
	(ii)	Data and revised conceptual models used in PA as part of the formal certification application for WIPP (Oct. 1996).
Grimsel	(i)	General confidence building in the dual-porosity concept for geosphere transport modelling (i.e. no significant processes overlooked) and in the transferability of sorption data from lab. to field.
	(ii)	No direct use of data from migration experiment (e.g. diffusion constants and depth of diffusion-accessible wall rock) in PNC and Nagra PA

Note: (1) The Opalinus-clay tests are still at the planning stage.

5. Where was the greatest success and the most significant failure?

Tests (1)	Greatest success	Most significant "failure"
Boom Clay	Close agreement between predictions based in lab. diffusion data and tracer test results on a larger scale has strengthened confidence in the PA migration model.	–
WIPP	Tests designed to test hypotheses and answer questions. Tracer tests have thereby contributed to the evolution of site conceptualisation.	Unsuccessful tests aimed at determining whether the effects of source-term complexity have been inappropriately attributed to matrix diffusion.
Grimsel	Enhanced confidence in dual-porosity model (but see "failure", opposite) and demonstration of consistency between lab. and *in situ* sorption values.	The differences in time scales between the Migration Experiment and PA mean that different phenomena may be relevant in the two cases (e.g. diffusion into low-porosity wallrock insignificant in Migration Experiment, but thought to be important PA retardation mechanism).

Note: (1) The Opalinus-clay tests are still at the planning stage.

6. How do you assess the results of the test in the light of its rationale and objectives?

Objective	Boom clay	WIPP	Grimsel
Understanding basic phenomena governing mobility of radionuclides.	Successful (although not cited as an objective): demonstrated that diffusion is the dominant transport mechanism.	Successful: demonstrated that transport is not limited to fractures. Tracers interact significantly with the matrix.	Successful: no new and safety-relevant phenomena identified - enhances confidence in understanding.
Direct determination of radionuclide migration parameters.	–	Successful: led to estimation of ranges for Culebra physical transport parameters.	–
Studying transferability of lab-data to *in situ* conditions.	Successful: demonstrated, on a scale of metres, the applicability of lab. diffusion data for tritiated water and iodine.	–	Successful: consistency of sorption data demonstrated for some weakly- and moderately-sorbing tracers.
Development and refinement of radionuclide transport models for PA.	–	Successful: led to refinement of processes included in transport model for PA.	–
Demonstration of predictability of radionuclide migration through a potential host rock.	Successful: conceptual model and data used for prediction of tracer tests applied in recent PA studies.	Successful: led to refinement and improved defensibility of conceptual model and parameters used in PA calculations.	Grimsel granodiorite not a potential host rock. However, "culture of rigorous and predictive model testing" has been established.
Enhancement of public acceptance.	Difficult to assess at the current stage.	Successful: based on presentations to date, public acceptance has been enhanced.	35 000 visitors to date at Grimsel Test Site. Effects on public acceptance difficult to assess.

7. Would you plan/design a new test in the same way now and what can be improved?

In Session II, this is only discussed in the context of the Grimsel Migration Experiment. It is acknowledged that several features of this experiment would be changed in hindsight: e.g. a more complete hydrological characterisation of the site and an earlier structural and petrological description of the flow path would be performed. Furthermore, greater involvement in the planning and design by performance assessors, at an early stage in the experiment, would have been desirable to ensure the production of PA-relevant data. In Session III, the question is discussed in the context of WIPP in the paper by MEIGS et al.

In addition to tests with injected, synthetic tracers, the paper by VIRA discusses the benefits of natural tracer studies. The potential advantages of such studies are:

- that they reflect transport in conditions and over time scales that are more relevant to PA than those that prevail in tests with injected tracers; and
- that they give information about rocks that would correspond to the near field of a repository – in several PA studies, the near field is key to the safety concept, rather than high-transmissivity water-conducting features of the far field.

VIRA concludes that "... we may have to live with the possibility of a leaking far field, but we should try to ensure as good as possible a near field." Further, "... Tracer tests are one possible means in site characterisation, but their application should be judged by their costs and benefits in relation to alternative methods and approaches."

2.3   Session III: Test Cases: Design, Modelling and Interpretation

The papers in Session III describe the experience and results of the completed or on-going tests at the URL (FROST et al.), at Äspö (the TRUE project; OLSSON & WINBERG), at WIPP (MEIGS et al.) and at El Berrocal (GUIMERÀ et al.). The tests are at different stages: tests at El Berrocal and WIPP are complete, tests at the URL are at an intermediate stage and tests at Äspö started only recently. The test programmes have many common aspects in the approaches adopted, in the results and in the conclusions. Only a few characteristic features of the test programmes could be pointed out in the summary presented here. Items presented for one programme could, in many cases, apply to other programmes as well. The questions set for Session III and addressed by the papers are summarised in the following tables.

1. What were the objectives of the test (model building and/or testing, hypothesis testing, methodological development, general understanding, demonstration and confidence building,...)? How were these objectives incorporated in the test design?

Tests	Objective	Incorporation in the test design
URL	To gain a better understanding of the processes affecting solute transport in fractured rock. Testing of the suitability of the porous-media equivalent method for fracture zones and moderately fractured rock.	Whole rock environment with three fracture domains (fracture zones, moderately and sparsely fractured rock) addressed at various scales (1-700 m)
Äspö TRUE	Development of understanding of radionuclide migration. Evaluation of the link between model concepts and realistic rock description. Assessment of applicability in site characterisation. Evaluation of usefulness and feasibility of modelling approaches. Provision of *in situ* data.	Test series of successively increasing complexity. Integration of experimental and modelling work. Predictive modelling and periodic evaluation of test results and successive improvement of models and test designs.
WIPP	Testing of important model features and recent hypotheses about transport. Quantitative estimation of important transport parameters. Demonstration of matrix diffusion. Determination of adequacy of fracture-matrix geometry, anisotropy and heterogeneity to explain results.	New improved test designs and equipment. More detailed characterisation of Culebra. Use of various tracers having different diffusion coefficients. Use of various pumping rates. Single well injection/ withdrawal tests. New modelling approaches.
El Berrocal	Development of methods for hydraulic characterisation, instrumentation development and data base generation. Integration of flow and transport research in heterogeneous domains: laboratory experiments, field work and modelling.	Pressure, temperature and concentration measurements in isolated borehole sections. Test methodology allowing versatile identification of flow and transport behaviour e.g. by use of dilution measurements both in natural and pumped conditions.

2. How does the test relate to the overall R&D programme (relationships to theoretical confirmation, to performance assessment, to public relations,...)?

Tests	Relation to overall R&D programme
URL	Evaluation of a concept for nuclear waste disposal. Documentation and demonstration of the feasibility of the disposal concept in an Environmental Impact Statement submitted for public, regulatory and scientific review.
Äspö TRUE	Generic demonstration of the function of the host rock as one barrier contributing to the multibarrier principle. Addressing the needs of PA by showing that pertinent transport data can be obtained from site characterisation or field experiments and that laboratory and *in situ* data can be related.
WIPP	Direct support for site specific PA. Confirmation of the conceptual model and parameters to be used in assessment.
El Berrocal	Generic hydraulic characterisation studies, building PA know-how and instrumentation development. Submission of test design plans to international scientific discussion and review.

3. *Which information was really obtained from tracer experiments performed till now (and which information was not obtained)?*

Tests	Information obtained	Information not obtained
URL	Adequate fluid flow models developed. Small and large-scale permeability variations within the fracture zones must be taken into account. Non-uniform transport properties needed to explain results fully.	Suitable transport model capable of simultaneously describing all the tracer tests within the same fracture zone.
Äspö TRUE	Hydraulic and other characterisation allowed the development of descriptive models that provide a basis for transport modelling. Predictions reasonably good for boreholes near to injection but uncertainties for more distant holes. Advection-dispersion data obtained but source term not optimal to study dispersion in detail.	Tracers from two of the four holes used for injection did not arrive (at least before test termination). Transport properties other that those related to advection (and to some extent to dispersion) not obtained.
WIPP	Refined conceptual model of Culebra. Heterogeneity of transport properties. Late-time slope (-5/2 instead of -3/2) of "SWIW" tests indicate a multirate diffusion model required.	Uniqueness of diffusion rate distributions, advective and diffusive porosities not demonstrated. Flow rates along the transport paths not known. Initial spatial distribution of the injected tracer slug not known.
El Berrocal	Data according to two alternative conceptual models: advection-dispersion or advection-dispersion-matrix diffusion. Flow rates through injection sections measured both without and with pumping. Most meaningful results obtained by models reflecting the 3D nature of the flow system.	Effects of slug injections not known and thus may be interfering with tailing due to other mechanisms e.g. matrix diffusion. Alternative explanations for the form of the break-through curves: (i) flow in fractures only, with diffusion into the matrix (ii) slow flow also within matrix, could not be distinguished.

4. *What is the reasoning behind the interpretation methodologies? Are alternative interpretation methodologies available? How to screen them?*

Tests	Reasoning	Alternative methodologies available and screening
URL	Success in interpreting previously performed tests (in fracture zones). Further evaluation of suitability of the porous-media-equivalent fluid flow and transport properties in fracture zones and in regions of moderately fractured rock.	Evaluation of other approaches such as discrete fracture network models. Tests at various scales and differently fractured domains. Radiometric analyses of fracture surfaces after completion of single fracture migration experiments.
Äspö TRUE	Parallel use of various interpretation methodologies (e.g. stochastic continuum and discrete fracture network) to interpret a series of tracer experiments with successively increasing complexity. Modelling in all phases of experiments: scoping, planning, pre-test, post-test, final evaluation.	All approaches and concepts used are checked continuously during the test programme. Injection of resin and excavation of tested rock volumes to reveal flow path geometries and tracer concentrations.
WIPP	Experience and results of interpretations of past tests. The need to demonstrate whether or not matrix diffusion is an effective phenomenon during transport in Culebra. The objective to distinguish between the effects of matrix diffusion and heterogeneity in permeability. Use of various conceptual models and checking against experimental results.	It is concluded that one must evaluate whether alternative conceptual models can explain the data. Various types of tests (e.g. RC and SWIW), together, are suited for providing insight into the important processes and for testing conceptual models. It is important to use various test-design features, like different pumping rates and injections into different locations.
El Berrocal	The interpretation is based on the conceptual model of fracture flow and solute transport. Two models were chosen and compared: radial advection-dispersion and radial advection-dispersion with matrix diffusion. The importance of accounting for experimental procedures, e.g. injection, in the modelling was emphasised.	Alternative models tested and the outcome of parameter values estimated. Simplified models led to parameters of doubtful validity for prediction purposes. Heterogeneity may be partly responsible for the observed results. Comparison of results with different tracers exclude the possibility that heterogeneity is the sole cause of the tailing.

5. What can tracer tests teach us about important transport mechanisms?

Tests	What can be learned about transport mechanisms
URL	Transport properties within fractured crystalline rock relevant to various scales are to be used in conceptual and numerical models of groundwater flow and solute transport.
Äspö TRUE	A better understanding of radionuclide transport and retention processes. Ability to obtain pertinent transport data from site characterisation or field experiments. Relation of laboratory data to retention data obtained *in situ*.
WIPP	Significant refinement of the conceptual model for transport at a site (Culebra). Occurrence of flow in various parts of a formation: in Culebra mainly within fractures and, to some extent, interparticle porosity and vugs connected by microfractures. Mechanisms, like multi-rate matrix diffusion, coupled to the flow.
El Berrocal	Models (together with test data) can be used to detect what kind of processes are important for the behaviour of solutes in the flowing groundwater. Diffusion into the crystalline rock was seen to be an important process.

6. How was spatial variability treated when conceptualising, designing, modelling and interpreting these tests (including simplification/abstraction steps when used in PA)?

Tests	Spatial variability treated
URL	Conceptually, the small-scale variability is averaged via the dispersion term. Large-scale variability is modelled numerically by taking into account differing thicknesses, permeabilities, porosities and orientations within flow and transport domains.
Äspö TRUE	The scale of the tests is within an interval of 1 - 100 m. In the tests, all scales of heterogeneity are accounted for in all phases of the tests. In single feature (fracture) tests, the aperture and property variation of the flow and transport paths is examined and, in block-scale tests, the variability of fracture properties and flow paths in the network is studied. Justification of simplifications for PA will be examined.
WIPP	The tests were specifically designed to reveal spatial variability and anisotropy in the Culebra formation. Important properties regarding inhomogeneities could be determined. These properties were accounted for in the modelling of the tests and in the derivation of the conceptual model for solute transport in the formation to be used in PA.
El Berrocal	Spatial variability was used as a concept in designing and interpretating the tests and also checked against simple (homogeneous) models. Experiments were performed so that, in addition to highly conductive fracture zones, the low permeable rock mass was tested. In the interpretation, this was explicitly taken into account.

7. How did performance assessors, modellers and experimentalists interact before and during the test? How did these interactions influence the test? How did the test design evolve?

Tests	Interactions
URL	Prior to any excavations and tracer experiments, a detailed site evaluation programme was carried out in co-operation with experimentalists and modellers from various fields. The tracer experiments are part of PA and, specifically, the prediction of potential radionuclide migration in plutonic rock bodies of the Canadian shield. The tests are, to a great extent, designed on the basis of PA needs and thus performance assessors have influenced the tests.
Äspö TRUE	A group of experts, "The Äspö Task Force on Modelling of Groundwater Flow and Transport of Solutes", has been engaged to provide advice on experimental design, predictive modelling and evaluation of experimental data. All experiments are based on thorough investigations performed at the site over about a decade. Interactions are encouraged and supported to a greater extent than is usual by organising the work into stages and iterative cycles.
WIPP	Based on site characterisation and a series of tracer tests performed in the 1980's, the outcome of the preliminary PA for the WIPP site was commented upon by numerous review and regulatory groups. The need to distinguish between alternative conceptual models was indicated. The recent tracer tests were carried out with extensive interaction between modellers and experimentalists prior to and during the tests. Based on the results of preliminary tests at H-19, additional testing was planned and performed at the H-11 hydropad and additional wells were drilled for further tests.
El Berrocal	A close link between theoreticians, modellers, lab and field experimentalists was established to evaluate existing experience on the design, performance and interpretation of tracer tests. Preliminary experimental designs were submitted for discussion and comments to an international group of experts. A multidisciplinary team was integrated into the work.

8. What are the limitations of tracer tests?

Tests	Limitations
URL	Not discussed explicitly from the point of view of individual transport mechanisms. At different scales, different limitations exist, but an averaging of small-scale heterogeneities seems to exist at transport scales over 30 metres (well dispersed break-through peaks). Further work is required in order to develop a model that can simulate all the tracer tests conducted during the different phases of tracer testing.
Äspö TRUE	According to the strategy of a staged approach, where different transport scales are addressed and the degree of complexity is successively increased, knowledge of important transport mechanisms can be obtained progressively. Thus, each tracer test has its limitations, but the next test can be based on experience from the previous one and more information gained (with new limitations). Integration with laboratory experiments is seen as crucial.
WIPP	The stepwise approach used was valuable for designing a good test (and gaining additional insight), but the approach could have been improved by adopting an even more evolutionary strategy over a longer time frame. Design of new tests would be integrated with ongoing laboratory programmes on rock diffusion and sorption properties. Tracer tests will always have the limitation that they cannot test the materials over the spatial and temporal scales of interest for PA calculations. Testing of alternative conceptual models is, therefore, essential.
El Berrocal	Some possibly remaining ambiguities in the modelling and interpretation of experimental results are discussed. More reliable results could be achieved when more realistic and complex models are used in interpretation. These models may, however, encounter difficulties with respect to software and CPU time and memory. It is important to incorporate realistic experimental conditions in the models (e.g. effects of the presence of the boreholes, natural groundwater flow and flushing during injection) which are often neglected.

All programmes appear to address the same problems of the groundwater flow and solute transport, but with somewhat different weightings. Strategies for design, modelling and interpretation of tracer tests are basically similar, even though the geological environments and media may be different. This means that it is possible to learn from the experience of all of these programmes for future experiments at other sites. Study of the experience and results of these programmes will be beneficial for any new tracer test programme.

2.4   Session IV: Aims and Design of Planned Field Tracer Experiments

In Session IV, three papers were presented. These provided descriptions of the planned experiments at the Kamaishi Mine (UCHIDA et al.), Palmottu (GUSTAFSSON et al.) and at the URL. Future tests to be performed at the URL were presented, along with on-going tests, in the paper by FROST et al. in Session III. This paper is therefore not summarised again here, although the questions for the Session IV address issues of aims and design from a slightly different perspective. The questions set for Session IV and addressed by the two remaining papers on future tracer-experiment programmes are summarised in the following tables. The two experiments are very different in scope, scales, environment and many other respects - e.g. Kamaishi experiments dealing with flow and transport in intermediate-scale fracture networks take place in an underground environment, whereas large-scale experiments at Palmottu are performed from the surface.

1. What are the new approaches to design, implement, model and interpret these tests?

Tests	Approaches
Kamaishi Mine	The focus on block-scale (10-100 m) flow heterogeneity and transport in a relatively tight rock described by the discrete fracture network concept and model is a relatively new approach, similar to that adopted by the URL and Äspö TRUE experiments (c.f. previous section). The experiments are integrated with laboratory experiments and natural analogue studies. The tests and their modelling have been designed to give more realistic and detailed information to be linked with PA geosphere transport models.
Palmottu	Combining pumping and tracer test, with simultaneous interpretation of drawdown and tracer break-through curves, is also a fairly new approach. The results will reveal properties of the present natural flow system at the natural analogue site, helping the forthcoming analogue studies on mobilisation and retardation of uranium in crystalline bedrock within and around the deposits.

2. What can be expected of further tracer tests (what is possible - what is not), what are their aims and how are they designed? What are the typical mistakes that need to be avoided?

Tests	Expectations and aims	Mistakes to be avoided
Kamaishi Mine	Tracer tests are used to derive flow porosity, dispersivity and connectivity information, and to test transport models with emphasis on hydrogeologic structure. By means of a discrete fracture network model, a realistic representation of heterogeneity at block scale will be achieved.	Not to expect, from short duration tests, information on slow, safety relevant processes, such as matrix diffusion. Not to underestimate the inherent limitations of tracer tests, such as non-uniqueness of results, in the sense that several processes may produce similar break-through curves. Not to perform tracer tests in poorly characterised structural and hydraulic environment.
Palmottu	The expected results of the combined hydraulic and tracer test will be hydraulic transmissivity and storativity values, dispersivities, flow porosity, leakage parameters, and possibly boundary parameters. If data so indicates, hydraulic anisotropy values may also be estimated.	To avoid ambiguities caused by uncertainties about hydraulic boundaries to the studied system. Not to interpret man-made artefacts (including equipment failures or effects of other activities nearby) and external disturbances as properties and phenomena of/in the studied transport system. A well kept "Log of Events" will help to avoid this.

3. What are the rationale and objectives of the planned test (model building and/or testing, hypothesis testing, methodological development, general understanding, demonstration and confidence building, ...)? How are these objectives incorporated in the test design?

Tests	Rationale and objectives	How incorporated
Kamaishi Mine	The rationale is to confirm the conclusion of the previous PA (H3) about the capability of the block-scale volume of the host rock to effectively retard radionuclide migration. The objectives are: 1) to obtain a conceptual model with realistic geometries and properties 2) to understand the hydraulic properties and geometries of barriers to flow. The results will be used: 1) as a basis for conceptual representation of the block scale in the next PA 2) to build confidence in application of the discrete fracture network model.	By detailed planning of the tests in a programme that advances in a step-wise manner. Preliminary testing preceding the main tests to optimise the chances of achieving the objectives. A well-defined hydraulic characterisation of the test site precedes the tracer tests. Inclusion of various scales and the study of various geometries and phenomena into the supporting test programme.
Palmottu	The rationale is to have more precise and quantitative data from a natural analogue study. There is a need to better constrain the boundary conditions of processes and events relevant to PA. This requires stricter physico-chemical constraints in natural analogue studies. The objectives are 1) validation of an updated conceptual hydrogeological model 2) identification of the main potential flowpaths at the scale of the test 3) increasing understanding about flow and solute transport properties at the site.	By using a combination of robust and well tested equipment and experimental methods/procedures, and making the design as simple as possible, without compromising the possibility of meeting the overall objectives of the experiment. By integrating quantitative and step-wise modelling throughout the sequence of tests, in order to optimise the experimental performance.

4. How does the test relate to the overall R&D programme (relationships to theoretical work, to laboratory experiments, to site assessment and confirmation, to performance assessment, to public relations, ...)?

Tests	Relation to R&D programme	Relation to site assessment or PA, PR
Kamaishi Mine	The experiments are part of an integrated programme of experimental activities to understand flow and transport in fractured rocks.	The results are used to develop and apply discrete fracture network models and concepts for PA. By applying a more realistic geosphere transport model, confidence in the result of the previous and future PAs will be enhanced.
Palmottu	The results give generic information on major flow and transport paths and migration behaviour of uranium and the members of its decay chains.	The experiments support PA work by means of quantitative study of the behaviour of uranium in a crystalline granitic bedrock environment (natural analogue).

5. How are lessons from previous tests integrated in the objectives/design of the planned test?

Tests	Lessons integrated
Kamaishi Mine	The tests can be seen as an extension of previous tests that have focused more on single features. No new methods are intended to be developed but experience from e.g. Stripa, Grimsel and URL will be drawn upon.
Palmottu	Experience from previous tests at the Hard Rock Laboratory at Äspö and El Berrocal site have been used in objective definition and test design. Equipment development and chosen experimental procedures are also in accordance with this experience and allows the optimal performance of the planned experiments.

6. What will be its contribution to confidence building in predictive modelling of radionuclide transport, to site characterisation/evaluation, and to performance assessment?

Tests	Contribution
Kamaishi Mine	It will contribute as part of an integrated programme including laboratory testing in single and multiple fractures, flow and transport tests in Japanese underground facilities, natural analogue studies, geological studies, and co-operative studies in international underground test facilities. It will support the development of numerical simulations of block-scale flow and transport, and will develop a link between discrete fracture network and geosphere PA models, thus supporting performance assessment.
Palmottu	It is an intermediate phase in the Palmottu analogue project and will be used by the project itself for further studies. The tests will make a contribution to confidence building in predictive modelling of radionuclide transport and PA in an indirect and generic way. The tests assess flow and transport in a subhorizontal fracture zone with intersecting vertical zones.

The tests summarised in this section are planned to be conducted in the near future (Kamaishi: starting in spring 1997 and to be completed by March 1998; Palmottu: to be conducted in early summer 1997) and the detailed designs may be changed.

3. RECORD OF WORKING GROUP CONCLUSIONS

3.1 Conclusions of Working Group 1: Practical Challenges
Chairmen: A. Hautojärvi, VTT (Finland) and H. von Maravic, EC

(a) Practical issues

Practical issues that were identified by Working Group 1 as those to be addressed during planning, performance and interpretation of field tracer experiments:

- A clear specification is required of hypotheses to be tested by the experiments. The number of unknowns should be minimised by performing the experiments on a system with:

  - well-defined hydraulic properties (a hydraulic definition of the transport pathways is desirable);
  - well-defined geological/structural properties;
  - well-defined geochemical properties.

- The limitations and uncertainties of field tracer experiments should be recognised; these are related to:

  - the experimental set-up;
  - scale (questions regarding extrapolation);
  - governing processes (the ability to distinguish between processes).

- Transferability of data (to a system of relevance to performance assessment) should be addressed in the analysis of field tracer experiments, bearing in mind:

  - scale (spatial and temporal);
  - geological differences;
  - flow conditions (e.g. forced flow vs. natural, unperturbed flow).

- Frequently, break-through curves are the only output of field tracer experiments from which to meet the goal of a meaningful and unique interpretation.

- Integration of information from many independent sources is most likely to lead to a general understanding of migration, e.g.:

  - tracer tests, combined with natural tracer tests;
  - field tracer test, combined with laboratory tests and natural analogue studies.

- Valuable data and support for performance assessment can be provided by field tracer experiments, irrespective of the performance assessment approach adopted (deterministic or probabilistic).

- Tests are more valuable if targeted at specific objectives, rather that "overall" tests aimed at many objectives: good questions produce good answers *(non specific questions produce hardly anything)*.

- Examples of objectives for good tracer tests are *(it should be recognised that other valid purposes exist)*:

    – pathway definition;
    – porosity-distribution definition;
    – investigation of chemical interaction;
    – model validation for performance assessment.

- The following undesirable characteristics should be avoided (some features of a good tracer test are given in the next section):

    – unknown, undefined transport pathways, geometry and boundary conditions;
    – unknown, undefined equipment behaviour (e.g. equipment failures);
    – unknown, undefined features of the design and artefacts;
    – unknown, undefined source term;
    – lack of pre-test predictions;
    – lack of reproducibility of results[2] (or lack of testing for reproducibility).

(b)   Features of good tracer tests

The following were identified as some characteristics of a well-designed tracer test (a fully general definition of a good tracer test was not considered to be possible):

- **Clearly stated rationale for the test, including**

    – identification of processes to be tested and why these processes are likely to be important for performance assessment calculations;
    – formulation of a well-defined conceptual model at the tracer test scale and at a larger scale that includes the effects of the heterogeneous nature of geologic media;
    – description of alternative conceptual models;
    – performance of pre-test calculations, based on alternative models.

- **A multi-disciplinary approach, with**

    – tests planned by a variety of technical expertise, considering (i) practical constraints; (ii) equipment artefacts and (iii) measurement limitations;
    – geochemical interactions (tracer-tracer and tracer-rock) taken into account;
    – hydraulic and geometric limitations taken into account.

- **A review of the design of the test, its results and its use in PA by outside experts.**

    – prior to the test, that the test, as designed, will provide important information on migration;
    – after the test, that the test results have provided important information on migration.

---

[2] Testing flow paths sometimes changes them, so reproducibility may not be possible.

(c)   Other issues

Other issues that were considered to be important are:

- The need for interaction with performance assessors and regulators:

  – A multi-disciplinary approach is advocated in which (i) a consistency of terminology and (ii) transparency of the relationship between terms and approaches should be striven for.

  – Interaction of experimentalists with performance assessment modellers should allow the *predictive* modelling to be tested.

- The value of an international review of the results of field tracer experiments.

(d)   Conclusions

Working Group 1 concluded that:

- the contribution of field tracer experiments to performance assessment depends on the geological medium and on the safety case (scenarios) considered,

- field tracer experiments can be designed and executed such that they provide valuable information to performance assessment.

(e)   Recommendations

The following recommendations were made by the Working Group 1:

- The purpose of a field tracer experiment needs to be well defined.

- Field tracer experiments will continue to be required for specific purposes, such as those outlined under "examples of objectives for good tracer tests", above.

- Further work is needed to tackle various practical/ technical issues.

- Continuation of an international review of the results of field tracer experiments is recommended.

3.2 Conclusions of Working Group 2: Rationale and Promises of Future Field Tracer Experiments
Chairmen: J. Vira, Posiva (Finland) and G. Volckaert, SCK/CEN (Belgium)

(a) Rationale: Historical perspective

• Typically in the past, there existed an overall need to demonstrate modelling capability and to "validate" performance assessment models (since relatively little experimental data were available about transport, this need was perceived both by implementors and regulators). This may have led to a degree of frustration, due to expectations being too high.

• Present trends are towards integrated testing strategies (incorporating lab, field and theory), iterative approaches and a perception, at least, of the need for performance analyst/modeller/ experimentalist interaction. The latter is in evidence in the trend towards integration committees, advisory groups, etc.

• A basic issue is the difference in the scales of time and space of relevance to the performance analyst (or "site characteriser") from those accessible to the experimentalist, which implies that abstractions are needed for performance assessment/site characterisation and tests cannot always be done in the way the performance analyst or the modeller would like to see them done.

• From the regulatory perspective, the view of Working Group 2 was that the implementer can hardly do without tracer tests, but, so far, no explicit requirements have been placed on them. Costs should be contrasted to the value of a licence; the regulator also needs support for their (licensing) decisions.

• A final point is that tradition may sometimes overrule rationale judgements. However, even if the scientific advantage of a test is not always clear, tracer tests may bring with them public acceptance benefits.

(b) Expectations: Practical objectives

The following practical objectives were identified by the Working Group 2:

• to support hydrogeological and flow modelling (e.g., to obtain information about connectivity, flow porosity, dispersivity);

• to describe transport pathways;

• to describe/measure interactions between rock, groundwater and moving tracers;

• to "validate" migration models (considered possible over the small spatial scales required for transport in clay);

• to build confidence in models and analyses and to increase scientific understanding;

• to improve testing methodologies;

• to increase public understanding and acceptance.

(c)   **Question of choice: How unique are tracer tests in providing the expected information?**

- Information on flow porosity/groundwater travel time would be hard to obtain by means other than tracer tests with *conservative* tracers.

- Information on interaction rates can be obtained using *weakly sorbing* tracers (but practical experiments are somewhat limited in scope).

(d)   **Critical challenges**

- How to perform the scale transformations needed for PA applications (from smaller to larger scales, for different boundary conditions):

  - Some processes can best be studied in small scale.

  - Could passive tests be designed even for fractured rock?

  - Long-term experiments may be possible during the active life-time of the repository.

- How to transport the information from lab to field, from site to site (critical for design of tracer tests in rock laboratories).

- How to create more sensitive test designs (basically, all the information from a tracer test is in the breakthrough curves - how much can be read from them and how can tests be made to discriminate processes and features?):

  - Better rock descriptions help focus on processes (fracture characterisation and classification).

- How to study slow processes such as matrix diffusion. How applicable is the Kd approach in interpretation of tracer tests?

(e)   **Broad recommendations**

- Build a strategy for transport analysis and modelling together – involve all parties who have related expertise – iterate – re-think.

- Find out relationships – study processes and interactions in the scale where you can get discriminatory information. "Validate" transformations to larger scales. Do not try to answer all the questions with a single test.

- Use independent information (as a consistency check).

- Develop testing with slightly sorbing tracers.

- Make use of international co-operation.

3.3 Conclusions of Working Group 3: Alternative Methods to Tracer Experiments

Chairmen: M. Heath, Earth Resources Centre, Univ. Exeter (United Kingdom) and W. R. Alexander, Univ. of Berne (Switzerland)

(a) Areas of discussion

Several areas were discussed within the Working Group 3:

- "traditional" tracer tests: within this heading came more or less all tracer or hydrological tests carried out to date.

- "realistic" tracer tests: this title was coined to describe those tests which utilise as near as possible the natural conditions of the rock body under investigation and, where possible, use relevant source terms. Although such tracer tests are not unusual in the environmental sciences, they are unknown (at least to the members of Working Group 3) in the radioactive waste disposal field.

- laboratory experiments: taken to mean any measurement of transport or retardation conducted in the laboratory.

- natural analogues: taken to include natural and archaeological analogues.

- natural tracers: in effect palaeogeohydrology and hydrochemistry.

- "anthropogenic" analogues: in other words, either accidental releases of material or man-made disturbances that are not yet old enough to be archaeological analogues.

The main conclusion of Working Group 3 was that the above noted alternatives were not alternatives *per se*, rather they should be viewed as complementary methods to the traditional tracer tests. Indeed, it was strongly felt that not only should traditional tracer tests be retained, but also that none of the alternatives could replace traditional tracer tests and that the best results were obtained when a true mix of techniques was employed.

Unfortunately, such mixes appear to be rare outside the Grimsel MI experiment, the WIPP programme and the on-going work at Kamaishi and it was felt that there is a much greater need for proper integration of such work in any planned projects. Clearly, the greatest advantage can be gained when such integrated planning is carried out from the first stages of any project.

A good example of the need for better integration was provided during the El Berrocal presentation during Session III, where it was pointed out that many complementary methods had indeed been applied, but the data had not been brought together.

A particularly keen discussion point was the suggestion (made during discussions during Session II) that tracer tests be employed during site characterisation. While several regulators at the meeting envisaged greater use of tracer tests throughout the characterisation (from during exploratory drilling to full facility excavation), concern was expressed that it would be difficult to carry out appropriate tests on appropriate features of concern at a particular site.

The most obvious problem relates to accessibility of conductive features (e.g. how can major conductive features be tested when the repository is designed to avoid them?) and experimental timescales (i.e. what point is there in carrying out site specific tests over a couple of years when the data required should be somehow representative of repository relevant timescales?). The conclusions were that, if something is to be done, then it should be done using natural tracers (i.e. more or less the standard palaeogeohydrological studies already carried out in many programmes), natural analogues (although no such relevant natural analogue is currently known and they probably will not be site specific) and realistic tracers (using natural groundwater gradients and doped repository materials such as cements and glasses over very long time periods).

As an alternative, it was noted that generic tracer tests still have much to offer when it comes to understanding specific performance assessment relevant processes and mechanisms. A good example of this is the investigation of the degree of confidence which can be placed in data produced in laboratory experiments. This has been tried in the Grimsel MI experiment with respect to applying laboratory Kd values to predict *in situ* retardation and comparing laboratory matrix diffusion data with natural analogue data - but this approach has been otherwise somewhat neglected.

Finally, it was pointed out that traditional tracer tests which do no more than match tracer output curves with some type of transport model were of little use as it is always possible to explain away a poor model fit by twiddling one or more of the numerous free parameters in the model. Conceptual models of geosphere transport and retardation can only be properly tested when the flow system is described in enough detail to remove as many of the free parameters as possible - otherwise, little confidence can be placed in the models.

(b)   Conclusions

1. There are no true alternatives to traditional tracer tests, but there are many complementary techniques which can be applied in conjunction with these tests.

2. Currently, these complementary techniques are not generally well integrated into traditional field tracer experiments.

3. Although greater use of field tracer experiments in site characterisation has been called for, this could be extremely problematic. If relevant features can be tested for relevant timescales, then the tests to be employed should be realistic tracer tests, natural analogues and the palaeogeohydrological studies already employed in many programmes rather than traditional tracer tests.

4. Generic tracer tests can still contribute much to the understanding of specific processes and the testing of how laboratory data may be transferred to the prediction of *in situ* retardation.

5. Describe the flow system - or little confidence can be placed in models supposedly tested by field tracer experiments.

3.4 Conclusions of Working Group 4: Integration of Data from Field Tracer Experiments into Performance Assessment

Chairmen: P. A. Smith, Safety Assessment Management Ltd. (United Kingdom), and P. Bogorinski, GRS (Germany)

The following points were agreed upon within Working Group 4 regarding performance assessment and the integration of data from field tracer experiments:

- **Performance assessment makes use of a combination of quantitative ("hard") and qualitative ("soft") information and analyses.**

The information and analyses, when taken together, should provide reasonable assurance that safety objectives are met. "Hard" information is, for example, data that can be input directly into calculational tools. "Soft" information is, for example, wide-ranging evidence that give confidence that safety assessment methodologies, models and data are appropriate.

- **The balance between "hard" and "soft" information depends on the approach chosen by the performance assessor, as well as regulatory aspects and cultural aspects.**

The factors are interrelated. The approach chosen by the assessor is dependent, in part, on the repository system (waste type, host-rock type, etc.), but is also influenced by the stage reached by the national waste management programme, by regulatory guidelines (whether time-frames are specified by regulations) and by the way in which implementors and regulators interact (cultural aspects).

- **Performance assessment is not necessarily about predicting reality; analyses are based on sets of assumptions and simplifications and it is recognised that judgement must be used in determining what simplifications and assumptions to make.**

For aspects of the repository system where current understanding is adequate, analyses generally aim at realistic prediction. However, for other aspects of the system, due to limitations in available information, the analyses are limited to conservative (over-estimates) of the consequences of the repository. A problem with this approach is that assumptions and simplifications are not always <u>unambiguously</u> conservative.

- **Field tracer experiments can serve to provide both "hard" and "soft" information for performance assessment.**

With respect to "hard" information (data), field tracer experiments are, in general, just one of a number of sources, complementing the information from laboratory tests, field hydrogeological tests, natural analogues and geological data. Limitations in the use field tracer experiments in acquiring "hard" data for performance assessment are that:

  - They are frequently carried out at "generic" locations (i.e. at sites not considered as candidates to host an actual repository), for example in order not to jeopardise the geological barrier through a very detailed characterisation process.

- The scales of space and time over which they are performed differ considerably from those relevant to performance assessment.

- They frequently focus on individual features (e.g. a single fracture or rock unit), rather than on the complete system.

Field tracer experiments are thus more widely used for identification of processes, testing of hypotheses, testing of the transferability of models and data between sites and confidence building and refinement of models. With respect to other "soft" information, field tracer experiments can test the applicability of laboratory data, build confidence in understanding of features and processes and in migration models and serve to gather and focus interdisciplinary expert input (for example, through discussion forums such as GEOTRAP). They can also stimulate the development of new experimental and analytical techniques and establish effective communication between field and laboratory experimentalists and modellers.

- **The contribution of field tracer experiments to performance assessment and the involvement of performance assessors in the design, modelling and implementation of field tracer experiments depend on the stage reached within the waste management project.**

It is clearly desirable that maximum information is extracted from field tracer experiments that is relevant to performance assessment and, to achieve this, interaction of experimentalists (in explaining what is feasible) and performance assessors (in explaining what is useful) is to be encouraged. The value of this interaction tends to increase as the project moves from acquisition of basic understanding of processes (e.g. the field tracer experiment programme at Äspö, Sweden), to the testing and refinement of performance assessment models (e.g. at WIPP, USA).

- **Field tracer experiments do not aim to provide analogues of actual repositories, with their associated large-scale heterogeneities and perturbations to geosphere conditions caused by the presence of the repository. They are currently (in most cases) restricted to tests that aim to address characteristics of the undisturbed geosphere.**

Field tracer experiments do not aim to address, in a single experiment, the full complexity of an actual repository system in terms of either structures or processes. Field tracer experiments do not need to be large scale to be useful (they are generally more successful for smaller-scale, simpler systems) and do not normally address alterations in geosphere conditions (scenarios) that might be caused by the presence of a repository: e.g. the effects on the geosphere of the high-pH plume from a cementitious repository. An exception is the investigation of excavation-disturbed zone properties using field tracer experiments reported by AECL. It is acknowledged that geosphere conditions are inevitably altered to some extent by artefacts associated with any field tracer experiment.

- **Field tracer experiments are most successful (in terms of "hard" information) for simple systems.**

The results of field tracer experiments are most easily incorporated into performance assessment (as data for models) when the system is simple in terms of the number of processes operating and in terms of structure (in particular, in the absence of fractures), as exemplified by the Boom Clay tests.

- **Tests on more complex systems (fractured media) can be difficult to interpret, but nevertheless modelling exercises frequently give "reasonable" results and may give support to certain performance assessment arguments.**

Compared to relatively homogeneous media such as Boom Clay, a larger number of processes operate in fractured media and characterisation is frequently incomplete. This can result in more complex break-through curves and alternative models that fit the curves equally well. Nevertheless, modelling exercises can be used to provide support for certain aspects of performance assessment models, such as the averaging of small-scale heterogeneities. Furthermore, it is not always necessary, for the purposes of performance assessment, to discriminate between alternative models, for example (i) if a conservative treatment is felt to be acceptable and (ii) if the feature/process concerned is, in any case, insignificant on the spatial and temporal scales of performance assessment.

- **It is rare in the modelling of field tracer experiments to examine conceptual model uncertainty in the sense of uncertainty in the <u>processes</u> that are operating. This suggests that a consensus may have been reached in the identification of processes relevant to tracer transport.**

Alternative models are applied to field tracer experiments, but these tend to be alternative ways of simplifying a geological interpretation of the system in order that transport modelling may be performed. Unexpected results are generally explained in terms of uncertainty and incompleteness of characterisation. Additional processes seldom need to be invoked. Sometimes it is not possible, on the basis of field tracer experiments and other information, to discriminate between alternative models. The different models may yield different predictions when applied over performance assessment scales of space and time. If this is the case, then the more conservative model is generally employed.

- **A continuum of experiments (in terms of spatial scale) exists between small-scale laboratory tracer (and other) tests, through larger-scale laboratory tests and smaller-scale field tests, to large-scale field tests.**

Smaller-scale tests offer better control of boundary conditions and the possibility of more complete characterisation. They are better suited to providing "hard" data for near-field performance assessment. Larger-scale tests offer the possibility of examining the operation of transport processes in larger-scale structures and may allow perturbations (e.g. of geochemical conditions) associated with sampling to be avoided. Such tests build confidence in the understanding of far-field processes. In general, the different types of experiments provide complementary information, that must be considered as a whole in performance assessment.

- **The following are suggested as components of a good tracer experiment programme:**

(i)   **A clear connection to performance assessment and site characterisation**

–   The aims of the tracer experiment programme should include a statement of the relevance of the test to particular aspects of performance assessment and/or site characterisation – performance assessment specialists should participate in the planning of the tests.

– The function of the tracer experiment programme within the broader national waste management programme should be indicated.

– The conclusions of the tracer experiment programme should indicate the relationship between the results and performance assessment and/or site characterisation.

(ii)   Identification of "success criteria" for the tests

– Success criteria should take due consideration of experimental errors.

– Means should be identified by which individual processes can be discriminated – for example, the unambiguous "signature" of a process in a break-through curve.

– In the particular case of matrix diffusion, "post-mortem" tests may provide a useful means by which to confirm the interpretation of experimental results.

An increased integration of field tracer experiments and performance assessment is noted in many national programmes. Field tracer experiments can, in certain cases, provide "hard" information for performance assessment. There is also, however, an increased appreciation of the qualitative aspects of performance assessment to which field tracer experiments can also contribute, providing evidence that the methodologies, model and data used in performance assessment are appropriate.

4.   SUMMARY OF FINAL DISCUSSIONS

- During the final discussion it was agreed that field tracer experiments provide useful information for performance assessment modelling. In particular, they allow testing between alternative conceptual models, and can demonstrate the adequacy (or inadequacy) of the understanding of the transferability of laboratory data to the field. Such discrimination between alternative models is essential as it is not always clear which is the more conservative. Moreover, field tracer experiments for sorbing tracers and for determination of the role of matrix diffusion are required by PA, and hydraulic characterisation of sites alone is insufficient for PA modelling.

- Despite their usefulness, there are a number of problems associated with field tracer experiments, not least that they are complex, expensive and take long periods of time to plan and perform. It is not possible, therefore, to carry out very many field tracer experiments. Moreover, the interpretation of the results of these tests is ambiguous, particularly if performed on a large scale.

- Cost-benefit analysis of tracer experiments is not easy because the end-results of such tests (in terms of data and understanding) is not predictable. The unexpected spin-offs of the test (unexpected tracer behaviour or unpredicted pathways, for example) are often the most valuable aspects of field tracer experiments but cannot be anticipated at the planning stage.

- Tracer experiments are normally on a small scale (metres), perhaps testing individual fractures, and the need was identified to develop conceptual models on the scale of the repository (hundreds to thousands of metres). The site-specific nature of tests and difficulty in extrapolating or transferring the results to other sites, particularly in view of rock mass heterogeneity, was also recognised.

- The possible role of field tracer experiments during site characterisation and development was discussed and it was concluded that, although good access to the rock mass might provide many opportunities for tracer experiments, it would be very difficult to identify the individual pathways and, therefore, to prove that the geosphere is not leaky with respect to fast channels. Furthermore, although it would be possible to analyse a few metres around emplacement tunnels and deposition holes, the results would not necessarily be representative of the rock mass as a whole if too few tests were carried out. Neither would the boundary conditions be very stable.

- Test-as-you-go tracer experiments during the excavation and operational period is consistent with the NEA concept of corroborative testing. They would be particularly relevant to fractured rock (in which flow periods would be reasonably short) but not in the Boom Clay (in which tracer movement would be extremely slow). As the operational period of a repository could be up to 100 years, it would be possible to carry out tests with relatively strongly sorbing tracers and to look at areas identified by PA to be of special concern. The identification of flow paths would not be possible and such tests would be confined to the investigation of processes within flow paths. Investigation of short circuits to the geosphere (e.g. through bulkheads or the excavation disturbed zone) would, however, be possible.

- Essential for the effective use of tracer tests is the narrowing down, as far as possible, of the unknowns addressed by a particular experiment. Also recognised is the need to understand the flow system to allow up-scaling.

- The use of gas as a tracer was suggested as a means of identification of fast pathways. The use of gaseous tracers could be particularly useful in clays, as gases generated in a repository in clay could create pathways for radionuclide migration. It was recognised, however, that the results of such tests would be difficult to interpret (due to two-phase flow) and could be ambiguous (fast pathways for gas are not necessarily the same as fast pathways for solutes). The results of a gas injection experiment in fractured slate in Cornwall were described as an example of a gas tracer yielding unexpected (but very useful) information on flow path distribution – see Lineham et al. Radiochimica Acta, 66/67, 757-764 (1994).

- Among causes for concern is the inconsistent use of terminology and assumptions, and the inappropriate use of data. Channelling in fractures, for example, is not treated consistently in models and different techniques are used for calculating aperture. Consistent assumptions must be used in different models that use the same parameter (i.e. the same assumptions should be made for, say, the transmissivity of channels throughout the analysis) and, for each parameter, the models/assumptions used to derive its value should be stated.

- Greater co-operation between experimentalists and performance assessment modellers, each representing different scientific cultures, would be especially helpful. To date, the PA input to field tracer experiments has generally proved to be difficult (as experienced at Äspö), while the input of these experiments to performance assessment is relatively minor in most cases. Great benefit could be derived from a more concise and clear presentation of PA-relevant experimental results and by involving PA modellers in the analysis of field tracer experiments results. It is an important challenge to bridge the gap between data users and data producers and to simplify the geological description for modelling purposes. While it is generally agreed that such communication is desirable, there are few cases where it has been unequivocally successful.

- Of particular importance to the future usefulness of field tracer experiments is their integration into overall programmes of investigation (laboratory and field) and PA model development. In this respect, it was proposed that an agreed strategy for integration should be implemented in order to ensure that each of the elements of a programme of investigation are properly co-ordinated.

PART B

WORKSHOP PROCEEDINGS

SESSION I

General Overview
Chairmen: C. Pescatore (NEA) and H. von Maravic (EC)

What Has Been Learned from Field Tracer Transport Experiments – A Critical Overview

Aimo Hautojärvi
VTT, Energy, Finland

Peter Andersson
Geosigma, Sweden

Geert Volckaert
SCK·CEN, Belgium

Abstract

The general rationale of performing tracer tests has been, and will probably be also in the future, to characterise properties of the medium in question regarding flow and transport of solutes to various degrees of detail. Beyond this common point several concepts and approaches have been and can be introduced. This leads into diversification of the concepts, terminology, modelling and parameters. To some extent the various branches may be equivalent or correspondence may be found between the entities used to describe the medium. On the other hand, two analogous concepts, e.g. a porous media versus fractured media, which can give equivalent results for the description of the hydraulic behaviour, may differ extremely regarding the transport behaviour.

A general concept is used in our paper as a starting point to address the need of a specific and detailed description of the actual flow geometry for obtaining the desired modelling parameters. We focus on experiments made in fractured media and the clay and salt alternatives for repository host formations are touched only cursorily, although some of the principles in general discussions apply also for these media.

In this paper we present an overview of lessons learned and give some examples of tests performed in different scales in fractured media. We discuss the possibilities for ambiguities and failures of tracer tests regarding the certainty to determine various transport mechanisms and properties as well as parameter values for modelling. We discuss also the reasons for possible deficiencies of the tests regarding the given objectives. A short introduction to various important elements of performing tracer tests, such as tracer injection, hydraulic conditions, multiplicity of the pathways and physical as well as chemical interactions, precedes the overview in order to fix the frame of discussion. Inherent uncertainties of field tests are brought into discussion to outline a scope of field tracer tests.

1. INTRODUCTION

Disposal of highly active radioactive waste into deep geologic formations is under planning and preparation in many countries. A principle of a multibarrier system is foreseen to prevent and reduce potential releases and migration of radioactive nuclides into the biosphere. The geosphere itself is the ultimate barrier in this chain. Increased interest and active research in the area of groundwater flow and solute migration has taken place both in the theoretical and model development side as well as in the experimental side in the last decade. Field experiments have been performed in different media at several sites in many countries. The results and conclusions drawn from these experiments have been and will be used in the performance assessment work needed to demonstrate the safety of the planned solutions.

International co-operation has been an essential part in many research programs offering a wealth of data and results to be discussed in order to compare experimental research, modelling and application to performance assessment approaches. This co-operation strengthens the basis on which the safety assessments have to be built. As now three years have passed since the last major international project INTRAVAL ended and new projects have been and will be launched it is a good time to review the outcome of field tracer tests and learn from the experiences to find out potential issues for improvement. Due to practical circumstances there always remain limitations, like financial, temporal, technical and in human resources, which bound the experimental possibilities. It is very important that the experimentalists and especially modellers, who may not follow the experimental phase of the project so closely, are aware of these limitations.

We will make an attempt in this paper to assess critically the outcome of performed field tracer tests in general and to overview the present status of knowledge on transport of solutes. The emphasis will be on fractured media where advection is a dominant process. Diffusion dominated cases are dealt with only cursorily. The examples come mainly from tests that were given for evaluation and modelling exercises in the recent international projects INTRAVAL and STRIPA.

In our paper we present an overview of lessons learnt with some examples of performed tracer tests in different scales in fractured media. We discuss the possibilities for ambiguities and failures of tracer tests regarding the certainty to determine various transport mechanisms and properties as well as parameter values for modelling. We discuss also the reasons for possible deficiencies of the tests regarding the given objectives. A short introduction to various important elements of performing tracer tests, such as tracer injection, hydraulic conditions, multiplicity of the pathways and physical as well as chemical interactions, precedes the overview in order to fix the frame of discussion. Inherent uncertainties of field tests are brought into discussion to outline a scope of field tracer tests.

2. RATIONALE – WHY HAVE FTTE'S BEEN PERFORMED?

Many of the Field Tracer Transport Experiments (FTTE's) in the nuclear waste management programs have been performed to give support directly or indirectly to the performance assessment in its various phases. In demonstrating the safety of a planned or constructed repository large rock volumes and long distances in the surrounding of the repository as well as very long time periods have to dealt with. This can not be based solely on experimental data but concepts and models have to be developed to extrapolate the results in space and time. The general understanding of geologic formations and hydraulic testing at a site can give an overall view on the water movement. Transport of solutes is, however, more dependent on certain details of the water flow paths than the pressure field and averaged flow rates. It is necessary, therefore, to study also these details in addition to the hydraulic testing.

To understand the transport times of non-sorbing solutes one has to know the flow porosity which normally is one of the parameter values that can be determined from tracer tests. To understand transport time distributions in heterogeneous media not only a single value but also a description of the geometrical nature of the porosity and possible spatial variability of the porosity is needed. This description is in practice possible only by means of theoretical concepts which may be based on media descriptions like porous or fractured and homogeneous or heterogeneous. The simplified picture could be e.g. packed beds, bundle of capillaries or set of fractures etc.

Field tracer transport experiments are needed to check the correctness (applicability, validity) of the concepts, theories, models and parameters used to describe the migration at the site in question, or in general in certain kind of media.

Usually there are also some other reasons, especially in early phases of waste management research programs, to perform FTTE's, like the need of learning and practising to perform and evaluate FTTE's or to test equipment.

It is clear that there is no single test that can solve all the problems together and at once but FTTE's are an ongoing activity in interaction with other elements in waste management programs. Actually, only small pieces of the whole puzzle can be put in place based on single tracer tests. Tracer tests are, however, very time and money consuming and it is tempting to foresee more outcome from the planned tests than is realistically possible. Overestimation of the capabilities of tracer tests to solve migration problems in general should be avoided and more emphasis should be given to the specific problems that are planned to be studied in a FTTE.

3. WHAT IS ESSENTIAL IN A TRANSPORT PROBLEM?

The transport of solutes through a distance within a medium is governed by the hydraulic conditions and water flow which may be seen at a very general level as streamlines going through the pore space of the medium. There are two important points to be discussed: 1) the geometry of the flow field and boundaries with medium, and 2) behaviour of solute molecules in this environment.

Let us take as an example one of the perhaps best known and "simple" situations, the transport of solutes diffusing in laminar steady flow through a straight tube. There we know the parabolic flow field. For impermeable, reflective boundary conditions at the pipe walls we know the behaviour of released solute plumes as a function of position and time. But it is not self-evident that the averaged concentration of the solute over the cross-section of the tube behaves like a moving Gaussian pulse spreading longitudinally in time with a constant dispersion coefficient. In fact, at "early times" it doesn't, and the dispersion coefficient approaches a constant value first after a while. This example illustrates the concept of streamlines and boundaries as a basis to solve the transport problem.

Thinking of the rather general result of transport behaviour of solutes in a flow system presented in the example above it is plausible that there may be (but not necessarily are) several concepts that produce the same results within a certain range of flow conditions in different kind of systems. One might ask: So what, does it make any difference what the system is like? For really non-interacting solutes it may not make any difference, but we are mostly interested in solutes that interact with the medium and then the differences in transport resulting from the various concepts of the medium may be huge.

As an example we may state that it is almost sure that it makes a big difference if the concept is based on e.g. packed beds or bricks, bundle of tubes, set of fractures or channels in fractures. Does it follow from this that we need to characterise every single cubic μm of the repository host rock in order to predict the transport?

4. SOLVING THE TRANSPORT PROBLEM

Speaking still in general terms, two things govern the behaviour and transport of the non-conservative (reactive) molecules

1) interaction rate with the boundaries of the medium versus longitudinal transport

2) behaviour in the medium outside the region of longitudinal transport.

The most interesting chemical and physical processes in the medium are usually sorption and molecular diffusion. Other processes may occur as well, like irreversible sorption and precipitation, but these are not usually accounted for because it is difficult to show that these processes take place in all possible conditions.

The problem can in principle be solved if the above mentioned two components of transport can be determined using a reasonably valid concept and parameters. These are needed to ensure the correctness of the extrapolation.

In a heterogeneous medium it would be difficult to determine these entities throughout the whole transport path for all potential transport paths, but fortunately only integrals (not local values) over macroscopic ranges of the two factors are needed, and small scale variations can be averaged out. The transport time consists of two additive factors. The behaviour of the non-interacting tracer is governed by the flow field alone (by definition). This contribution is added to the transport time of reactive tracers spent at the surfaces of the medium or in the medium. Conservative (non-sorbing) tracers can interact with the medium in the sense that it may diffuse through the boundaries and spend some time outside the "flow region". In this picture we count e.g. stagnant areas of flow, which are directly connected to the flow field, to the "flow region". The flow field contribution to the transport time is usually not so interesting from the point of view of performance assessment but, it is important when experiments are interpreted.

The interaction rate versus longitudinal transport can be determined either by direct measurements (e.g. by comparing transport of sorbing and non-sorbing tracers) or theoretically with the help of concepts and models. The behaviour in the medium can be revealed by laboratory and in-situ measurements supported by modelling.

5. TRANSPORT EQUATION: VELOCITY, DISPERSION, SORPTION, DIFFUSION INTO THE MATRIX

The widely accepted terminology and formulation of transport in fractured media includes usually the processes: advection, dispersion, sorption (on fracture surfaces and in the matrix) and diffusion into the matrix. The theoretical formulation starts often from the parallel plate concept but it can be generalised e.g. to a heterogeneous, variable aperture case quite easily.

There are some problems applying this stream tube concept for other than point sources. For a point source one stream tube, where no significant variation of properties in the transverse direction of the stream tube exist, is a reasonable approximation. For larger source term extents some questions arise. How large could the extent of one stream tube be? Does dispersion account for different velocities? Is the dispersion really Fickian? How is sorption and diffusion through the boundaries coupled to the stream tube? What is the interaction between stream tubes?

These points should be addressed in the modelling and evaluating process of a FTTE, and when extrapolating the modelling to repository performance assessment (PA). A direct application of the transport model to reproduce experimental results, without a transparent description of the underlying flow geometry, is not satisfactory. This is not, however, an easy task because unfortunately the experimental data does not usually allow one to distinguish between the different flow concepts – in the future something must be done to improve this situation.

Noticing all these difficulties one may end up to ask if the usually presented formulation and the corresponding quantities are the right ones to go with? The answer can be yes, but their role has to be understood correctly and generally enough. There are relations between them and e.g. locally wildly varying quantities like velocity and fracture aperture produce together a very stable quantity: their product is related to the flow rate which is to some extent "invariant" along the transport path in given flow conditions.

The flow rate as the important quantity gives a very strong connection to the hydraulic characterisation of a site which is "easier" to perform than direct characterisation of transport properties. The hope is that with a valid flow field concept and necessary hydraulic measurements the transport at a site could be under control, and only some FTTE's would be needed to ensure and demonstrate this.

6. WHAT ARE TRACER TESTS?

Basically tracer tests mean that transport properties in an unknown domain in a medium are studied by sending a signal through the domain. Comparing the source signal and the registered output signal one should deduce which processes have been active during the transport and which kind of transport path geometry was involved. This is, of course, a very demanding task, and any possible support from other measurements and observations is needed for the interpretation.

In the evaluation it should be realised and accounted for that the input has certain spatial and temporal characteristics which are reflected then in the output signal. It is obvious that the input has to be known for a sensible analysis of the system. The significance of the source term is unfortunately underestimated too often.

For a reliable evaluation of the tracer test the flow field should optimally be in steady state, and if disturbed by the experimental procedures themselves the disturbances should at least be well controlled and known. Hardly any system can be studied by just one single measurement, rather a series of measurements is needed with different conditions. The greater number of processes has to be studied the more measurements are needed.

7. WHAT HAVE TRACER TESTS TAUGHT US?

In homogeneous media and especially when transport is diffusion dominated, the control of performing and evaluation of tracer tests is easier in the sense that much less ambiguities regarding e.g. the input or source term exist than in heterogeneous media. The experimental conditions are much closer to the ideal theoretical behaviour. The situation is most difficult with advection dominated, unknown heterogeneous flow systems.

In spite of extensive efforts to characterise experimental sites, most tracer experiments in fractured media have to be performed in the latter conditions. We have learned in international modelling exercises that in the case of fractured media there are often more concepts than experiments, and that it is usually not possible to distinguish between the different concepts and models.

Some of the reasons for this are that it has not been possible to control and measure the source term well enough for purposes of process identification. It is also known that two or more processes can produce the same kind of behaviour of the output signal. Even if the behaviour of two processes is different, it is difficult and uncertain to extract the effects of the two processes from the combined result. The flow fields are not always constant introducing uncertainties and ambiguities. Unknown processes or flow conditions may be responsible for the fact that the tracer recovery is often significantly less than 100 %.

Taking all these uncertainties into account, it can be concluded that it is rather easy to obtain a reliable median or mean transport time and the variance of the transport times, but other characteristics of the break-through curves are then already questionable. In the course of performing and analysing the tracer tests in the past we have learned what are the main weak points in the FTTE's performed so far, and know now better how to tackle the geosphere transport problems in a more efficient and accurate way.

The lessons learnt include that we have learned to take conceptual and analytical uncertainties into account, and that we appreciate more and more the role of "predictive" modelling in testing our ability to understand transport processes and bedrock features affecting transport. We have gained an increased knowledge and experience to organise a FTTE together with geologists, hydrologists, modellers, experimentalists, and performance assessors to the interactive and iterative effort of solving the transport problem.

8. SHORT REVIEW OF FTTE'S

This review concentrates in some characteristic details of some FTTE's performed recently. The review is by no means a comprehensive summary of complete results or conclusions, rather it aims at pointing out some important features of the tests that are worth discussing when preparing for future tests. Many valuable results were obtained, and there are many things that have been learnt from these tests, but there are also points which may be rethought and improved in the future. These and other performed FTTE's are valuable material to be studied carefully when the overall potential of getting information through tracer tests is assessed. Comprehensive reviews on tracer transport tests have been presented recently, e.g. by Gelhar et al. [1] mainly on the dispersion problem, and by Andersson [2] more related to the nuclear waste disposal.

Finnsjön radially converging and dipole tests

A set of tracer tests in a gently dipping fracture zone in crystalline rock at Finnsjön was performed. The objective was to study and determine transport phenomena and parameters in major fracture zones and to use the results for calibration and verification of radionuclide transport models. An additional objective was to develop and improve experimental equipment and methods [3]. The tests were handled as one test case (Case 5) in the INTRAVAL project in its both phases. Already in the early planning of the INTRAVAL project, the importance of interaction between experimentalists and modellers as well as of the predictive calculations were emphasised. It turned out, however, that in many of the INTRAVAL cases the tests were already performed before modellers could comment the test plans. The Finnsjön case was an exception in this sense, although the schedule was very tight so that comments and predictions could be presented but modellers' further contribution to the tests was limited. In spite of quite comprehensive test program, the experimental results and data were insufficient to distinguish between disparate models. Tracers were injected by circulating and mixing the tracer in the borehole section, but sampling of the concentration was not frequent enough to have an accurately determined source term. The extent to which matrix diffusion occurred in the experiments remained unclear.

Stripa 3-D

Groundwater flow and tracer transport through a three dimensional block of rock above a drift excavated for that purpose was studied in the Sripa mine. These tests formed the INTRAVAL Case 4. The rationale was to understand and quantify transport processes relevant to the safety of a final repository for high level radioactive waste. The measured water flow distribution studied over more than 700 m^2 was observed to be very uneven. Tracers were injected at nine different points and 167 break-through curves of six different tracers were measured [4]. The tracers were injected with overpressure. This may have spread the tracer in the vicinity of the injection point in an unknown way. Due to on-going activities in the mine disturbances occurred during the tests. These facts made it difficult to extract the effects of various transport phenomena, including matrix diffusion, from the results.

WIPP-2

In the second phase of the INTRAVAL project, the simulation of flow by means of stochastic 2-D modelling approaches was studied using the available extensive hydraulic data. In addition to that results of tracer tests (mainly from Hydropad 11) performed at the site were modelled. Anisotropy of the Culebra formation was seen to have an effect on the transport, but some uncertainty remained about the role of heterogeneity. The diffusion from the fractures of the dolomite formation into the highly porous matrix had a significant effect during transport according to the modelled results [5,6]. The parameters could not, however, be determined unambiguously because the effects of uneven flow distribution and source term spreading due to slug injection and borehole flushing were not known.

Stripa SCV

Tracer test in the block scale was performed in the Site Characterization and Validation project in the Sripa mine. The aim was to study water flow and tracer migration in a fracture zone as well as in the rock outside the zone. Compared to the Stripa 3-D experiments the injection technique had been changed to produce a constant injection flow rate instead of constant pressure. Care was taken that no major disturbances due to other activities would occur. The injection took place with slight overpressure. The resulting break-through curves were significantly smoother than in the Stripa 3-D case and agreement between modelled and measured results was better. Dyes and metal complexes were used as tracers, and they showed different break-through curves even though injected simultaneously. This difference could be explained with a small difference in their K_d values meaning that dyes would be slightly sorbing. The sorption was so weak that it could not be seen in laboratory measurements. This sorption could have enhanced the effect of matrix diffusion to the extent that the maximum values of the break-through curves differed roughly by a factor of two. It was concluded that flow in the fracture zone is channelled, and similar to the channelling in the average fractured rock [7].

VLJ-RT

Tracer tests were performed in the VLJ Research Tunnel at Olkiluoto, Finland. The tests were run between two boreholes being 6 metres apart. One of the 56 mm holes was bored at the location of a planned full scale simulation hole. In the first phase the test was run between the 56 mm holes and later from the 56 mm into the 1.5 m diameter hole. Geological and fracture mapping of the cores and later observations from the 1.5 m hole revealed that the rock was tight and sparsely fractured. Single fractures could be identified being responsible of hydraulic connections. The flow through the injection section was measured carefully with high time resolution to know the source term accurately. The experimental set-up in the later tests allowed to change the water of the injection section without disturbing the flow through the section. The relatively short injection pulse allowed a better analysis of the tail of the break-through curve. It was concluded that effects of matrix diffusion can be seen first when the not necessarily Fickian hydrodynamic dispersion behaviour is known accurately [8].

9. FTTE SCALES, PURPOSES, CONTROL OF FLOWS AND SOURCE TERMS

Different aspects of migration can be studied in different tracer tests. There is no single test that would solve all the transport problems. An optimal combination of test scales and types should be found out to gain as much as possible knowledge about transport of solutes and, ultimately radionuclides in various kind of media. The practical constrains for the tests, set by the environmental conditions in the field, should be analysed in test planning. The degree of achievable accuracy usually increases towards smaller scales. More sophisticated tracer tests to study transport phenomena can be performed better in smaller that larger scales. Large scale studies are essential for understanding of water flow in that scale and to determine parameters describing the flow field. The larger the scale is, the longer are usually also the test times and repetitions, and changes of test conditions are not possible to a great extent. A general idea of relations between scales, purposes of studies, governing flow fields and importance and possibilities to control the source term is presented in Table 1.

scale	purpose of study	flow field	source term
regional > 1000 m	conn	natural	not important
site 100 m - 1000 m	conn + por + disp	natural + pump	somewhat contr
block 10 m - 100 m	conn + por + disp + trans	pump + bkg	contr
detailed 0 m - 10 m	por + disp + trans + sorp + md	pump (var) + bkg	adj pulse

conn=connectivity, por=flow porosity including channelling, disp=dispersivity, trans=other transport properties, md=matrix diffusion, natural=natural flow, pump=pumping, bkg=background flow, (var)=various rates, contr=controlled, adj pulse=adjusted short pulse

Table 1. Relations between scales, purposes, flow conditions and source term controlling in tracer tests

The importance of various transport phenomena in migration from repository towards biosphere at different scales should be assessed by means of PA methodology and reflected in FTTE's.

10. ADDRESSED QUESTIONS

The Programme Committee has prepared a list of questions to be answered in this paper. The background for the answers to these questions is presented in the previous chapters and direct answers will be given here more explicitly. It should be remembered, however, that it is certainly impossible to cover all of the performed tracer tests with one general answer. The answers should be seen to represent more a trend among the tests than being applicable as such to any individual experiment.

1) Was the rationale of these tests clear enough?

The rationale for each test has been clear, but often too optimistic and unrealistically wide taking into account the available resources for fulfilling the goals (belonging to the rationale in question). For porous media the typical rationale is the need to know the flow and transport porosity together with the dispersion in one, two, or three dimensions. It is a simple and clearly stated rationale. Required test arrangements and procedures may be quite complicated, though. Experiments in fractured media have been reasoned often in similar terms than those in porous media. This may be a severe problem if the analogy does not hold. A characteristic phenomenon of fracture flow is channelling, and the rationale of many experiments has been based on this. Experimentally it is a challenge to address channelling in undisturbed rock and natural-like flow conditions.

2) Which information were really obtained? What have field tracer tests taught us about important transport mechanisms?

Flow velocities and dispersion can usually be obtained quite reliably and accurately. Beyond that the obtained information depends much on the used concepts and modelling. Usually, there are at the same time many different concepts and models that can explain the results. There are thus ambiguities in the interpretation, which can not be resolved due to lack of experimental data. The governing transport mechanisms can not be distinguished in such cases. The most debated and perhaps also most important transport mechanism is the matrix diffusion. It is extremely difficult, if not impossible, to show the effect of matrix diffusion on the break-through curves in field tracer transport experiments. It is certainly not enough to fit an advection-dispersion-matrix diffusion model to a BT-curve and extract the various transport mechanisms from the model parameters.

3) What use was made of the so obtained information?

The information has strengthened mostly our understanding on the flow in the media, and the transport calculations in performance assessments are based directly or indirectly on the flow characteristics. Still, there are discrepancies in the basis of transport modelling as partly discussed under question 2).

4) Have they helped to build confidence in the predictive modelling of radionuclide transport for performance assessment purposes?

The tracer tests have partly helped to build confidence in the transport modelling. There are important gaps to be filled before a satisfactory level of confidence can be achieved.

5) Where are the failures? Were these failures clearly reported? What are the lessons learnt from them?

The failures of fulfilling the goals of the tests regarding understanding of transport mechanisms, were mainly in test limitations and partly also in unfavourable test procedures for the study of transport mechanisms. The reports emphasise usually a good agreement of the experimental and modelled results, and possible ambiguities are not assessed critically. In this sense the "failures" or insufficiencies of the tests are rather hidden than clearly reported. The lessons learned from the failures of these kind are that more and better tests and detailed characterisation of the test site is needed before the transport mechanisms can be revealed and studied in field.

6) What can be expected from future field tracer tests?

In the future tests, the ambiguities will be reduced and tests will be more specialised in transport mechanism studies compared to the "overall" type of tests made in the past. This may mean various tests in various scales and combining of the results from different tests on site, generic and laboratory tests. Different concepts and models are distinguishable when compared with the test results. It seems that not all of the tests can be performed in field at a specific site. The tests at a disposal site will be even more limited in number. Characterisation of hydraulics is the main task at a disposal site and the performance assessment has to rely on the relations between the hydraulic and transport properties studied at other places, possibly nearby, and even on generic studies.

REFERENCES

[1]   L.W. Gelhar, C. Welty, and K.R. Rehfeldt, A critical review of data on field-scale dispersion in aquifers, Water Resources Research 28 (7), 1992 pp. 1955-1974.

[2]   P. Andersson, Compilation of tracer tests in fractured rock, SKB Progress Report 25-95-05, Swedish Nuclear Fuel and Waste Management Company (SKB), January 1995, Stockholm.

[3]   The International INTRAVAL Project, Phase 1, Summary Report, December 1993, The Coordinating Group of the INTRAVAL Project, Swedish Nuclear Power Inspectorate (SKI) and Nuclear Energy Agency, Organisation for Economic Co-operation and Development (OECD/NEA), Paris, France.

[4]   H. Abelin, L. Birgersson, J. Gidlund, L. Moreno, I. Neretnieks, H. Widén, T. Ågren, 3-D Migration Experiment - Report 3 Part I, Performed Experiments, Results and Evaluation, November 1987, Technical Report STRIPA PROJECT 87-21, An Organisation for Economic Co-operation and Development, Nuclear Energy Agency (OECD/NEA) project managed by Swedish Nuclear Fuel and Waste Management Company (SKB), Stockholm, Sweden.

[5]   T.L. Jones, V.A. Kelley, J.F. Pickens, D.T. Upton, R.L. Beauheim, P.B. Davies, Integration of Interpretation Results of Tracer Tests Performed in the Culebra Dolomite at the Waste Isolation Pilot Plant Site, Sandia Report Sand92-1579-UC-721, August 1992, Sandia National Laboratories, Albuquerque, New Mexico, USA.

[6]   The International INTRAVAL Project, Final Results, The Coordinating Group of the INTRAVAL Project, Swedish Nuclear Power Inspectorate (SKI) and Nuclear Energy Agency, Organisation for Economic Co-operation and Development (OECD/NEA) 1996, Paris, France.

[7]   L. Birgersson, H. Widén, T, Ågren, I. Neretnieks, L. Moreno, Site Characterization and Validation - Tracer Migration Experiment in the Validation Drift, Report 2, Part 1: Performed Experiments, Results and Evaluation, January 1992, Technical Report STRIPA PROJECT 92-03, An Organisation for Economic Co-operation and Development, Nuclear Energy Agency (OECD/NEA) project managed by Swedish Nuclear Fuel and Waste Management Company (SKB), Stockholm, Sweden.

[6]   A. Hautojärvi, M. Ilvonen, T. Vieno, P. Viitanen, Hydraulic and Tracer Experiments in the TVO Research Tunnel 1993-1994, April 1995, Report YJT-95-04, Nuclear Waste Commission of Finnish Power Companies (YJT), Helsinki, Finland.

The Contribution of Field Tracer Transport Experiments to Repository Performance Assessment

P. A. Smith
Safety Assessment Management Limited, U.K.

P. Zuidema
Nagra, Switzerland

Abstract

Models of solute transport in the geosphere are, in general, derived from hypotheses concerning:

- the types of structures present in rock
  (possibly supported by direct observations in tunnel walls, cores, etc.),

- the transport processes that convey the tracers within the relevant structures
  (supported by current understanding of geosphere transport),

- the rates and spatial extent over which these processes operate
  (supported by independent field and laboratory experiments – e.g. batch sorption and laboratory diffusion experiments).

Field tracer-transport tests can be used to provide support for individual hypotheses, for the overall models and for the methodologies to derive parameter values for the models, and thus to build confidence in the applicability of the models and data used in performance assessment. The present paper describes performance assessment needs with respect, for example, to confidence building. The types of confidence building, as well as other information, that can be obtained through the process of modelling the results of tracer tests are outlined. The value of predictive modelling is compared to that of "inverse modelling". The different ways in which the results of tracer tests can be applied in performance assessment are outlined, both where the rock in which the tests are performed is similar to a potential host rock and also where there are significant differences. In spite of the importance of tracer tests, there are limitations in the information that they can provide, particularly in the understanding of slow processes and processes operating over long times and large distances. These limitations are discussed.

1. Introduction: Performance Assessment and the use of Models

1.1 The components of performance assessment

An assessment of the performance of a radioactive-waste repository comprises the following three basic components:

1. An **evaluation** of the evolution of the repository system;

The evaluation must be quantitative, but, because of the long time scales over which the evaluation is required, cannot be based in direct observations. Rather, the evaluation relies on:

- A scenario analysis, in which a set of scenarios, representing alternative concepts for the future evolution of the repository system, is derived from a comprehensive list of features, events and processes (FEPs). Some or all of these scenarios are selected for quantitative, consequence analysis.

- A consequence analysis, in which (i), the structures within the repository and its environment, (ii), the relevant processes operating within these structures and (iii), the rates and spatial extents over which these processes operate, are described quantitatively in a set of models. Due to uncertainty in these descriptions, the model "predictions" do not necessarily aim at realism, but can rather be bounding: for well-understood aspects of the repository system, the model descriptions aim at realism and, for less well-understood aspects, conservative, simplifying assumptions are made that aim to over-estimate adverse consequences of the repository. Ranges/distributions of parameter values and alternative model assumptions typically need to be considered.

2. **Building confidence** that the "predictions" of consequence analysis are sufficiently reliable;

This involves building confidence:

- that the list of FEPs identified within the scenario analysis includes all safety-relevant phenomena, that the interactions between FEPs are adequately represented and that an adequate set of scenarios has been identified, covering uncertainty in the evolution of the repository system,

- that the models, data and computational tools are adequate and that uncertainties are taken into account, either in the set of models and ranges/distributions of parameter values selected for the analyses or in conservative assumptions.

3. **Assessment of available information**

This typically involves a discussion of the meaning of the results (for example, in terms of compliance with regulations), an evaluation of uncertainties and a statement of confidence in the results, in the light of various confidence-building measures (validation).

1.2   The role of models in performance assessment

Because of the need for quantitative evaluation of the performance of the repository system over long time-scales, the use of models is central to performance assessment. The repository system as a whole is commonly described by a chain of assessment models that each relate to a particular component of the system (e.g., a near-field release and transport model, a geosphere-transport model and a biosphere model), with a series of supporting models and hypotheses that serve to translate field and laboratory data into assessment-model input parameters (e.g., a hydrogeological model to translate the results of borehole tests and observations to geosphere-model input parameters such as flow-wetted surface and Darcy flux). An example of the relationship between supporting models and hypotheses and an assessment model is illustrated, in Figure 1. The figure is based on the geosphere-transport modelling performed in recent Swiss performance assessments [1], [2].

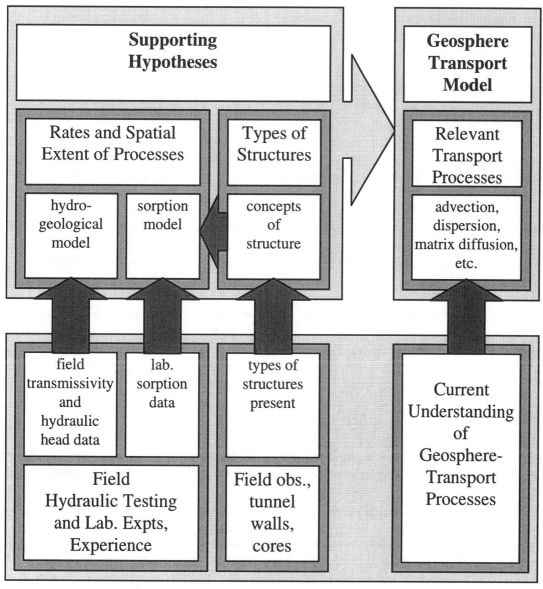

**Figure 1. Relationship between supporting models and hypotheses
and a geosphere-transport model**

An assessment model is often a (conservatively) simplified version of a more detailed research model, with the research model aiming at realism rather that bounding predictions. The same types of relationship illustrated in Figure 1 also apply to research models. The greater level of detail of the research models can arise because:

- Research models are designed for use on systems that are more fully characterised than is achievable in practice at a real site. A research model may be applied to only a (particularly well-characterised) part of the system to which an assessment model is applied. In the case of geosphere transport in a fractured medium, a research model may be applied to a tracer test in an individual, particularly well-characterised fracture. Detailed characterisation is likely to be impracticable for the fracture-networks of relevance to performance assessment (Section 4.2).

- The need to perform large numbers of calculations in performance assessment, in order to explore parameter sensitivity and to estimate the consequences of uncertainty, may require an assessment model that is relatively simple to avoid excessive demands on computer time and memory. The same constraints do not generally apply to research models.

Research models are based on a full representation of performance-relevant features, within the current understanding of the system to which they relate, and provide the theoretical framework on which assessment models are built. In particular,

- Research models provide a tool for validation, testing the hypotheses concerning structures, processes and rates/spatial extents by allowing model predictions to be compared with the results of laboratory and field experiments.

- Research models provide a tool for decisions on simplifications in the formulation of assessment models. Having confirmed understanding of the system, a model developer is in a better position to decide where simplifications can be justified in terms of either conservatism or insignificant effects.

In the context of this discussion of the use of models in performance assessment, a third class of models is mentioned briefly. These models are simplified analysis tools, based on a further simplification of the assessment models. They are fast to run (e.g. analytic solutions), can be of use for sensitivity analysis, allowing a wide region of parameter-space to be covered comprehensively (e.g. [3]), and can assist in the interpretation of assessment model results (e.g. [1]).

1.3   The components of a geosphere transport model

As illustrated in Figure 1, a model of geosphere transport, whether a research model, an assessment model or a simplified analysis tool, is, in general, derived from hypotheses concerning:

- The types of structures present in the rock

This includes the identification, classification and geometrical description of structures (fault zones, fractures, channels, etc.). Hypotheses concerning structure typically involve extrapolations of direct observations in the field, in tunnel walls, cores, etc., possibly supported by other evidence (e.g. geophysical studies).

- The transport processes that convey the radionuclides (in the case of an assessment model) or tracers (in the case of a research model) within the relevant structures.

This includes solute transport processes, such as advection, diffusion and sorption, but may also include a range of other processes including precipitation at reaction fronts, colloid-facilitated transport and gaseous-phase transport. These hypotheses are supported by current understanding of geosphere transport.

- The rates and spatial extent over which these processes operate.

This includes the parameter values that quantify the transport processes, e.g. Darcy velocity and, for fractured media, flow-wetted surface, diffusion coefficients and porosity distribution. These hypotheses are supported either by field experiments by or independent laboratory experiments – e.g. batch sorption and laboratory diffusion experiments.

2. Performance–Assessment Needs

Performance assessment needs, with respect to the prediction of radionuclide transport in the geosphere, where field tracer tests can provide a contribution, are in the following areas:

2.1 Confidence building and identification of uncertainties

The types of confidence building that can be addressed by field tracer transport tests are:

- Confidence that relevant structures and processes have been identified and that a satisfactory methodology exists by means of which rates and spatial extents of structures and processes are determined from laboratory and field data.

The success of a model, that is consistent with available information about the experimental system, in reproducing the results of an experiment builds confidence that relevant features and processes have been identified and that the methodology (e.g. supporting models) for quantifying rates and spatial extents is satisfactory.

With regard to uncertainties:

- Identification of remaining uncertainties in the understanding of structures and processes and reduction of these uncertainties.

The success of alternative models, also consistent with available information, in reproducing the results of an experiment indicates the degree of uncertainty in the understanding of these features, processes and rates/ spatial extents. The failure of alternative models serves to falsify hypotheses concerning structures, processes and rates/ spatial extents and thus narrows the degree of uncertainty.

The use of field tracer transport tests for confidence building and identification of uncertainties is further discussed in Section 4.1.

2.2   Assessment model formulation

If the research models need to be simplified for performance-assessment purposes, the understanding of tracer tests by means of these models can ensure that, in the assessment models

- key structures and processes are represented in the assessment models,

- where omissions/ simplifications are made, predictions do not err on the non-conservative side,

- laboratory and field data are used in an appropriate manner.

Field tracer tests can also be used to "calibrate" a model, either for use in further tests or, more rarely, for direct use in performance assessment. The direct use of tracer tests in the characterisation of a site is discussed further in Section 4.2.

3.   The Modelling of Field Tracer Tests

As indicated in Section 2, when a field tracer test is performed, it is the process of trying to understand the result, often by means of a (research) model that attempts to reproduce the experimental results that contributes to performance assessment needs. A model can be tested for its ability to reproduce the results of field tracer transport tests in two broad ways:

1. Predictive modelling, where predictions are generated by the model using parameter values derived from independent measurements.

To obtain maximum benefit from predictive modelling, it is desirable to make predictions in a transparent manner before the tracer test is performed, with a clear methodology defined for the setting of parameter values. It is also desirable to establish "success criteria" for the predictions, taking full account of experimental errors.

2. "Inverse modelling" where parameter values are adjusted until a best fit of the experiments are found.

The type of confidence building that is provided by inverse modelling, though valuable, is generally less convincing than that provided by predictive modelling. It is necessary to recourse to inverse modelling where independent measurements are either not available or unsuitable for the determination of all parameter values. It is, however, essential, from the point of view of confidence building, to show that the fitted parameter values are at least physically reasonable and consistent with any available independent information.

Both types of modelling can, by demonstration of "goodness of fit" (within the success criteria), build confidence that the relevant features and processes have been identified and incorporated in a model that is used either directly, or in simplified form, in performance assessment, i.e. that no key processes have been overlooked. It is important to examine as many plausible alternative models as possible in order, by falsifying some of these alternatives by their failure to predict experimental results, to narrow down the range of conceptual model uncertainty and to identify the processes that are most important.

A significant difference between predictive modelling and inverse modelling is that:

- predictive modelling can build confidence in the methodology for deriving rates and spatial extents from field observations and from independent field and laboratory data and to show that parameter values obtained by these means are acceptable.

- inverse modelling can be used to "calibrate" a model, providing parameter values for subsequent application in predictive modelling of further tracer tests (i.e. running the same system in different modes - e.g. with different flow rates and with different tracers) and, in some circumstances, of a repository system in performance assessment.

The use of tracer test results in performance assessment, both for confidence building and model calibration, is discussed in the following section.

4. Use of Tracer-Test Results in Performance Assessment

4.1 Confidence building and identification of uncertainties at "generic" sites

Field tracer transport experiments have, in several cases, been performed at locations that are not intended as sites for future radioactive waste repositories. Tests at such "generic" sites can be useful:

- in providing experience in the practicalities of obtaining field data and relevant laboratory data and in stimulating the development and refinement of measuring devices (e.g. for hydrogeological testing, laboratory sorption tests and diffusion experiments),

- in establishing successful communications between geologists, laboratory and field experimentalists and modellers,

and, of more direct relevance to performance assessment, in various types of confidence building discussed in Section 2.1. Specifically:

- in building confidence, by a "trial application", that the methodology for the construction of a transport model from these and other data is appropriate,

- in building confidence that relevant structures and processes are understood (although structural details are likely to be site-specific) and that no relevant processes have been overlooked,

- in building confidence that laboratory data can be correctly applied.

The success of models in predicting the results of a tracer test (or, preferably, a sequence of many tests with the same system run in different modes) have, in the case of fractured, hard rock (e.g. [4]):

- illustrated the importance of an understanding of small-scale geological structure in constructing a model that, in performance assessment, can be confidently expected either to be realistic or to err on the side of conservatism,

- demonstrated the existence of the matrix-diffusion phenomenon and its key importance as a retardation phenomenon in performance assessment and, more generally, demonstrated that current understanding of geosphere-transport processes is sufficient to understand transport on the scale of these tests,

- demonstrated that, with appropriate consideration of the differences in conditions, laboratory data (e.g. on sorption and diffusion) can be applied to field-scale experiments, giving confidence that they can be applied in performance assessment.

In order to address the topic of conceptual-model uncertainty, in the international INTRAVAL project, seven different teams applying different model concepts attempted to model radially-converging and dipole tracer experiments at the Finnsjön site in Sweden [5]. All conceptual approaches were found to fit the experimental results reasonably well, indicating the range of uncertainty in the understanding of structures, processes and rates/ spatial extents. Conceptual-model uncertainty is of importance to performance assessment if it leads to uncertainties in predictions made for the relevant temporal scales, which are considerably greater than those characterising tracer tests. Predictions made using the different approaches were found to diverge considerably over these larger time-scales and it was concluded that it is not possible to extrapolate a (research) model calibrated on the experimental scale to simulate a much longer time-scale case. As discussed in Section 2, assessment models are thus based on (conservative) simplifications of such research models.

It is more difficult to cite examples where the failure of alternative models has served to falsify hypotheses concerning structures, processes and rates/ spatial extents and thus to narrow the degree of uncertainty. To some extent, this may be because a consensus has been reached as to the broad processes that are relevant to transport (e.g., in the case of fractured media: advection, dispersion, matrix diffusion, sorption). However, there is also an unfortunate tendency to emphasise (in the literature) instances where models are successful in reproducing experimental results, rather than where alternative models have failed.

4.2   Direct use of tracer tests in the characterisation of a site

Where field tracer tests are performed in a geological medium that is a potential host for a radioactive waste repository, "inverse modelling" of the results can, in principle, provide data for direct use in performance assessment. Key data that could be obtained from tracer tests are, for example,

- flow porosity,

- retardation data (e.g. sorption Kd-values),

and, for fractured media,

- flow-wetted surface,

- the distance to which radionuclides can diffuse from fractures into adjacent wallrock.

Problem areas in such direct use of field tracer transport tests in the characterisation of an actual site are the heterogeneity of the site (Section 5), which may mean that many such tests are required for characterisation, and the practicalities of performing large numbers of tests without, in the process, perturbing the favourable hydrogeological properties of the site. It is, therefore, not routine practice to perform field tracer tests to obtain data as part of the characterisation of an actual site. Such tests have, however, been used to support and refine models of important geological features in the region of a site [6], [7]. Furthermore, a related type of experiment, tracer-dilution tests, has been used, for example, in as part of the characterisation of the crystalline basement of Northern Switzerland [8], [9]. Dilution tests involve the flushing of a section of a borehole where there is known to be a water-conducting feature with a tracer and monitoring the tracer concentration as a function of time. These tests are potentially useful, since they minimise the perturbation of the natural hydraulic gradient (conventional hydraulic tests deliberately perturb the gradient) and can be used to infer flowrates. However, practical difficulties and experimental artefacts can limit their usefulness and, in [8], [9], they were used to bound, rather than predict a realistic value for, groundwater flowrates.

5.   Treatment of Spatial Heterogeneity – Limitations of Field Tracer Tests

The successful modelling of field tracer tests indicates that heterogeneity on a smaller scale than the tests themselves can either be averaged out (as in the case of heterogeneous sorption) or can be treated in a simple manner (e.g. the modelling of hydrodynamic dispersion as a diffusion-like process); this is also an important assumption underlying most geosphere assessment models. However, a major difficulty with field tracer tests is that they operate in domains of space and time that are significantly different to those of relevance to performance assessment: no information is provided on processes that, though irrelevant on the spatial and temporal scales of field tracer tests, may be important over scales relevant to performance assessment - i.e. slow processes and processes operating on large-scale features, such as:

- matrix diffusion into fracture wall rock (it is the effects of diffusion in loosely consolidated fracture infill material that are apparent in some field tracer tests, e.g. at the Grimsel test site, Switzerland [4]),

- hydrodynamic dispersion in extensive fracture networks.

The identification of slow processes may be regarded the domain, for example, of natural-analogue studies, rather than of field tracer tests. In the case of large-scale spatial heterogeneity, attempts have been made to perform and model tracer tests on extensively fractured rock masses, e.g. in Phase 2 of the international Stripa Project in Sweden ("Migration in a Large Fractured Granitic Rock Mass") [10] and in tracer tests at the Fanay-Augères Uranium Mine in France [11].

In the case of the large-scale experiment at the Stripa Mine, alternative models were applied, each representing a single realisation of the network, and it was concluded that "... the collection of transport models, based on various assumptions ... was not able adequately to reflect features of the tracer breakthrough curves". This inability of any of the alternative models to reproduce experimental results was attributed to:

- uncertainties in the geometries of flow paths through the network,

- uncertainties in the geometrical features of channels.

i.e. a far more detailed characterisation would be required in order to model realistically transport through a fracture network.

In the case of the tracer tests at the Fanay-Augères Uranium Mine, a different approach was adopted. No attempt was made to reproduce the detailed features of the breakthrough curves. Multiple realisations were generated of the network and the ensemble of model predictions compared with broad features such as the overall duration of breakthrough and the recovery rates. Consistency between the ensemble of predictions and observations was found, but the range of predictions within the ensemble was large and thus, perhaps, of limited value to confidence building (e.g. a range of predicted recovery rates of 2-70%, with observations, at 4 measurement points, of 5%, 6%, 14% and 45%). Again, more specific predictions would require a more (possibly unattainably) detailed characterisation.

6. Discussion and Conclusions

This paper has describes performance assessment needs with respect to various types of confidence building and data acquisition. Many of these needs can be addressed by the process of trying to understand the results of field tracer tests in terms of models. These are, typically, detailed research models that are then used as a basis for simplified assessment models. In general, predictive modelling is more valuable than "inverse modelling" in terms of confidence building. Maximum benefit to performance assessment from predictive modelling of field tracer-transport tests can be obtained by:

- making predictions in a transport manner before a test is performed, which involves:

  - a clear methodology or strategy for setting free parameter values (from, for example, independent field and laboratory experiments),

  - establishment of criteria for "successful prediction",

- examining as many plausible alternative models as possible, which involves:

  - covering the full range of conceptual-model uncertainty,

  - aiming to falsify alternatives,

  - publicising "failed" model alternatives and not only those that give successful predictions,

  - taking account of all available information (including independent experiments and general, scientific knowledge and experience about structures and processes),

  - aiming to identify the most important structures and processes,

- extrapolating results to the larger spatial and temporal scales that are relevant to performance assessment, to provide an indication of the consequences of conceptual-model uncertainty.

Apart from practical limitations and perturbations caused to the experimental system by the tests themselves, a fundamental limitation of field tracer-transport tests is in the scales of space and time that they can address. Large-scale structural heterogeneity, in particular, is a highly relevant phenomenon in geosphere performance assessment. The geosphere provides an effective barrier to the transport of a radionuclide if the transport time through the geosphere exceeds the half life. Even where this is true on average, if a few pathways exist where the transport time is less than the half life, then these can dominate the performance of the geosphere barrier as a whole. Large-scale heterogeneity is generally treated simplistically in performance assessments. As discussed in this paper, it is not necessary in performance assessment to predict actual transport behaviour, but rather to make predictions that are confidently expected to over-estimate the consequences of transport. It is thus possible, to some extent, to compensate for lack of information on large-scale heterogeneity through the use of conservative assumptions. The difficulty with this approach, however, is that credit for the performance of the geological transport barrier may be reduced to such a degree that it plays only a minor role in repository safety; this will be the case, for example, if the existence of a "fast channel" through the host rock, that affects releases from a sufficiently large part of the repository inventory, cannot be excluded. In these circumstances, the information from field tracer tests, interesting as it may be, is of rather limited use to performance assessment.

7. References

[1]   Nagra, Kristallin-I Safety Assessment Report, Nagra Technical Report Series, 1994, NTB 93-22, Nagra, Wettingen, Switzerland.

[2]   Nagra, Bericht zur Langzeitsicherheit des Endlagers SMA am Standort Wellenberg, Nagra Technical Report Series, 1994, NTB 94-06, Nagra, Wettingen, Switzerland.

[3]   PNC, Research and Development on Geological Disposal of High-Level Radioactive Waste, First Progress Report, 1992, PNC, Tokyo, Japan.

[4]   Alexander, W. R., McKinley, I. G., Frick, U. and Ota, K., The Grimsel Tracer Field Tracer Migration Experiment - What have we Learned after a Decade of Intensive Work?, 1996, these proceedings.

[5]   Andersson, P. and Winberg, A., Conclusions from WG-2: The Analysis of the Finnsjön Experiments, Procedings of an NEA/SKI Symposium: GEOVAL '94, Validation through Model Testing, Paris, France, 1994, 87-99.

[6]   Beauheim, R. L., Meigs, L. C. and Davies, P. B., Rationale for the H-19 and H-11 Tracer Tests at the WIPP Site, 1996, these proceedings.

[7]   Meigs, L. C., Beauheim, R. L. and McCord, J. T., Design, Modelling, and Current Interpretations of the H-19 and H-11 Tracer Tests at the WIPP Site, 1996, these proceedings.

[8]   McNiesh, J. A., Andrews, R. W. and Vomvoris, S., Interpretation of the Tracer Testing Conducted in the Leuggern Borehole, 1990, NTB 89-27, Nagra, Wettingen, Switzerland.

[9]   Spane, F. A. Jr., Description and Results of Tracer Tests Conducted for a Deep Fracture Zone within Granitic Rock at the Leuggern Borehole, 1990, NTB 90-05, Nagra, Wettingen, Switzerland.

[10]   Gnirk, P., OECD/NEA International Stripa Project, Overview Volume II, Natural Barriers, 1993, SKB, Stockholm, Sweden.

[11]   Cacas, M. C., Ledoux, E., de Marsily, G., Barbreau, A., Calmels, P., Gaillard, B. and Margritta, R., Modelling Fracture Flow with a Stochastic Discrete Fracture Model: Calibration and Validation to the Transport Model, Water Resources Research, 1990, Vol. 23, no. 3, 491-500.

Regulator's Point of View on the Use and Relevance of Field Tracer Transport Experiments

P. Bogorinski, B. Baltes
GRS (Cologne, Germany)

1. Introduction

Nuclear waste arising from the operation of nuclear power plants as well as from the use of radionulcides in medical treatment, industrial applications and at research facilities has to be disposed of in a way that future generations will bear no risk to their health and to the environment. Therefore most countries plan to built repositories within deep underground geological formations. In the course of licensing procedures compliance with regulations has to be demonstrated by means of safety analyses for the operational as well as for the post closure phase.

The goal of post closure safety analyses is to demonstrate with a high degree of confidence that in case of a potential release of radionuclides from the repository the consequences to man and to the environment are below regulatory limits. These analyses have to be carried out for long time periods after sealing of the facility. The most important issue is whether numerical safety analyses take into account those migration pathways which may provide the fastest return for the radionuclides to the biosphere and result in the highest calculated exposure.

For the regulator who has to review long-term safety assessments during a licensing procedure for a nuclear waste repository questions arise

— how valid the predictions of the future evolution of the facility and the site are,

— how good the models used to carry out these predictions represent the nature of the site,

— which experimental evidence is needed to support the hydrogeological model of a site on which analyses are based,

— to which extent site investigations have to be carried out to provide the information needed as input into the numerical models, and

— how experiments can be designed to confirm the results of the safety assessment.

In this context we would like to discuss in our paper based on the experience gained during assessments of repositories sites in Germany whether field tracer transport experiments are a sensible tool that may be used to gain confidence in the safety evaluation of the facility.

2. The German Waste Disposal Programme

Before we discuss the relevance of tracer tests for performance assessment we give a brief overview of the situation of waste disposal in Germany. Three waste disposal facilities are currently under consideration.

– The disused Asse salt mine had been operated as an underground research facility which included the experimental disposal of low and medium-level wastes.

– The Gorleben salt dome had been selected to host a final repository for all types of waste. It is currently under site characterisation including the construction of an underground exploratory mine.

– The disused Konrad iron ore mine had been selected to host a final repository for non-heat-generating wastes. The facility is currently in the licensing process.

– The disused Bartelsleben salt mine near the village of Morsleben is in operation as a repository for low-level radioactive waste.

All sites are characterised by a sequence of sedimentary layers above the host rock ranging from jurassic and cretaceous formations as in the case of Konrad to quaternary sandy and clayey deposits as in the case of Gorleben.

3. Performance Assessment Modelling of Sedimentary Systems

In sedimentary systems two types of heterogeneities are to be taken into account, firstly the different geological formations at a scale of a few metres to some hundreds of metres vertically and hundreds of metres to kilometres horizontally. Heterogeneities at this scale may be modelled explicitly for groundwater and nuclide transport simulations. However, these formations themselves are not homogeneous. Mineralogy may vary at very small scales of a few centimetres causing spatial variations of hydrological and transport parameters and these variations may change over the whole region. It is obvious that modelling heterogeneities at this scale explicitly for groundwater and transport simulations with the numerical tools available is not feasible to date since it would require a huge computational effort both in terms of time and memory. As it is state-of-the-art now properties have to be averaged over larger volumes. In doing so potential connectivities existing in low permeable rocks between highly conductive zones of the aquifer system may not be taken into account in the model. However, the question is whether such connectivities are relevant for the results of a long-term safety assessment.

In contrast to hard rock the porosity of sediments is relatively large which means that the storativity for radionuclides is relatively large as well as the dilution capacity. This is emphasised by the results of groundwater flow and transport analyses, where travel times range in thousands to hundred thousands for non-reactive or even millions of years for reactive nuclides. In addition for most sedimentary rocks present at suitable sites the retardation properties are more favourable than in hard rocks.

Therefore, if such connectivities are really small they might well represent a potential fast path for radionuclides to return to the biosphere but only for such a small fraction that no relevant contribution to exposure will result.

For the Gorleben repository project long-term safety assessment calculation have been carried out in the frame of the CEC EVEREST-project /CEC 96/. The geology of the site is characterised by a glacial erosion channel which cuts through the tertiary clay cover of the salt dome as far down as to the cap rock. This channel has been filled afterwards by quaternary deposits consisting of a sequence of sandy/gravely and clayey/silty layers. The Lauenburg clay complex covers most of the area of interest separating a lower aquifer which is in direct contact with the salt formation from an upper near-surface aquifer. This clay layer contributes largely to the isolation capacity of the geosphere at the site. Here the crucial question is whether gaps exists in this clay layer which connect the lower and the upper aquifers thus providing a potential migration pathway for radionuclides escaping from the salt formation into the biosphere.

In the Konrad long-term safety assessment /BfS 94/ travel times of non-sorbing radionuclides from the repository to the biosphere have been calculated to be in the order of 300,000 years, assuming conservative estimates for the parameters governing the radionuclide migration. In this base case simulation the main migration pathway is along the host formation itself for a distance of approximately 30 kilometres from the repository to the biosphere. Additionally several variants of geosphere migration simulations had been carried out to account for uncertainties in the hydrogeologic characterisation of the site. These variants addressed mainly the permeability of clayey layers above the repository. However, travel times range in the same order of magnitude even in cases where relatively high values are estimated for clay permeabilities and radionuclides penetrate the clay barrier directly above the facility with a path length in the range of a few kilometres.

Fault zones within a sedimentary system may be modelled explicitly with their own hydrogeological and transport properties assigned to them. However, such detailed modelling will not be needed for individual fractures which may appear as single features within these faults zones. In contrast to hard rocks the surrounding sedimentary soft rocks will have a sufficient retardation capacity accessible by diffusion that the impact of single fractures on radionuclide migration is negligible.

4. Use of Tracer Tests in Charactering Heterogeneities

One issue to be discussed is how tracer tests represent the real situation of a repository site. At first we have to consider the temporal and spatial scale in long-term safety assessments. In time it ranges from a few thousands to some hundreds of thousands of years and in space from some thousand metres vertically to several tens of kilometres horizontally. It is quite clear that tracer tests cannot cover these scales.

Usually tracer tests will only cover a small area of the region of interest. In order to characterise the geosphere at a repository site one would have to measure the distribution of an injected tracer within the system. This would require a large and dense population of boreholes which would disturb the system by introducing artificially heterogeneities. If novel methods would be available to measure tracer concentrations within the system from the surface without disturbing it by boreholes the results from such investigations will be more useful.

Nevertheless, tracer tests on small scales in space and time are a sensible tool to determine hydrogeological and transport parameters such as permeabilities, porosity, and retardation coefficients at selected locations, in particular where alterations due to geologic events such as e.g. faulting may be suspected.

Next issue is the depth of the repository which is in the range of some hundred metres. Conditions near the surface are not necessarily representative for those deep underground. Two options are available for carrying out experiments at the potential repository level, namely deep drilling or an underground laboratory. Deep drilling require considerable effort which would be only worthwhile if a representative location could be found whereas the construction of an underground laboratory will disturb the hydrogeological and potentially the geochemical conditions and therefore will not represent the real site conditions for the time periods after closure of the facility. Nevertheless results from such experiments might be useful to characterise the immediate vicinity of the facility.

5. Conclusions

Tracer test carried out at an underground laboratory or within geologic formations similar to those at a selected site for waste disposal might be very useful to validate the generic capabilities of models to be used in groundwater flow and nuclide transport simulations for long-term safety assessments and thus might enhance the confidence in performance assessments.

Furthermore, tracer tests might also be very useful to acquire site specific data such as porosities, diffusivities or retardation properties that are relevant for modelling the migration of radionuclides through the rocks of the geological barrier.

In addition tracer tests may be used to characterise the immediate vicinity of the underground facility, in particular the excavation damaged zone.

However, to date we would not request specific tracer tests for the sole purpose of characterising the hydrogeologic and geochemical conditions at the site of a proposed nuclear waste repository at a scale relevant for performance assessment.

The fundamental issue in soft sedimentary rocks is absence of distinctive flow paths such as fractures where water would move much faster than in the rock matrix. Therefore, tracers injected into such a groundwater flow system disperse to a much larger extent and travel much longer than in fractured hard rocks. Recovery of tracers in monitoring boreholes would be poor and breakthrough would be difficult to interpret at a large scale. Therefore, field tracer transport experiments are in our view not the appropriate means to characterise the heterogeneities of a sedimentary system.

References

BfS 94   The Konrad Repository Project - From an iron ore mine to a repository for radioactive wastes, Salzgitter 1994

CEC 96   European Commission Evaluation of Elements Responsible for the Effective Engaged Dose Rates Associated with the Final Storage of Radioactive Waste: EVERST Project, Volume 3a: Salt formation, site in Germany in print.

SESSION II

Rationale Behind Field Tracer Experiments

Chairmen: M. Heath (Earth Resources Centre, United Kingdom)
and P. Bogorinski (GRS, Germany)

Field Tracer Experiments in Clays

G. Volckaert (SCK·CEN, Mol, Belgium)
A. Gautschi (NAGRA, Wettingen, Switzerland)

1   Introduction

The objectives and rationale behind the design of field tracer tests depend strongly on the type of rock but also on the type of site and on the state of advancement of the waste disposal research programme. Here the objectives and rationale of field tracer tests for two different clays, types of site and different project stages are discussed. We discuss tests performed in the Boom Clay at Mol, Belgium and tests planned to be performed in the Opalinus Clay in north-west Switzerland within the framework of the international Mt. Terri Project.

The Boom Clay is a plastic Tertiary clay. The Mol site is a potential repository site and associated research on it was started at the end of the seventies. The tracer tests are performed in the HADES underground research facility of the Belgian Nuclear Research Centre (SCK·CEN) which has been in operation since 1984.

In the Mt. Terri Project, the Opalinus Clay, a well-consolidated Middle Jurassic shale (claystone) formation is being studied in a service tunnel of the Mt. Terri motorway tunnel that cuts through an anticlinal structure in the Jura mountains. The Opalinus Clay is currently under investigation by the Swiss National Cooperative for the Disposal of Radioactive Waste (Nagra) as a potential host rock for high-level radioactive waste. A potential siting region has been selected in north-east Switzerland, where the Opalinus Clay is in a relatively undisturbed tectonic environment. First scoping studies on the Opalinus Clay were started at the end of the eighties and the international Mt. Terri Project started with the excavation of niches and a first drilling campaign in early 1996. The field tracer test will start in 1997 after a series of laboratory tests (feasibility study).

2   Main characteristics of the considered clays

2.1   The Boom Clay

The Boom Clay formation of north-east Belgium is a marine deposit of Rupelian (Middle Oligocene) age, i.e. 30 to 35 million years. The unit consists predominantly of intimately mixed clay and silt and minor sand. Bedding is mainly defined by rhythmic decimetre-scale variations in mean grain size (Van Echelpoel and Weedon 1990). The carbonate-rich levels contain widely spaced septarian limestone nodules (Vandenberghe and Laga 1986). Some beds also contain pyrite concretions and/or important fractions of organics. Although the Boom Clay is of marine origin, its porewater is dominated by sodium bicarbonate. At Mol, the burial depth of the Boom Clay layer is 180 m and its thickness is about 100 m. From a hydrological viewpoint the Boom Clay is an aquitard with very low hydraulic conductivity and from geomechanical viewpoint it is an overconsolidated plastic clay. The mineral composition and some important hydro-mechanical properties of the Boom Clay are given in Table 1.

2.2 The Opalinus Clay

The Opalinus Clay of Northern Switzerland is a marine shale (claystone) formation of Middle Jurassic age (Lower Dogger, mainly Aalenian, i.e. 180-190 million years). The formation - named after the ammonite *Leioceras opalinum* - consists of well-consolidated, dark grey, micaceous shales, partly with thin sandy lenses, limestone concretions or siderite nodules. Based on its clay, sand and carbonate content, the Opalinus Clay can be subdivided into several litho-stratigraphic units. The average mineral composition and some important hydro-mechanical properties are listed in Table 1. Hydraulic tests in deep boreholes and hydrogeological maps from a total of 6400 m of tunnel sections in the Opalinus Clay of the Jura mountains indicate that the formation has a very low hydraulic conductivity, although joints and faults were present in the sections studied (Gautschi 1996). Porewaters of the Opalinus Clay are of Na-Cl type with a total dissolved solids content of 20 g/l at the Mt. Terri site (Gautschi et al. 1993).

Tectonically, the Mt. Terri site is situated in the southern limb of the Mt.Terri anticline, dipping to the south with 30 to 50°. Here, the Opalinus Clay has a thickness of 150 m and an overburden of roughly 300 m. The overall tectonisation of this part is rather weak, but detailed mapping of the tunnel walls clearly revealed the presence of numerous minor faults, which can be divided into thrust and normal faults (Geotechnical Institute Ltd. 1995).

Table 1 Mineralogy and some important hydro-mechanical properties of the considered clays (NEA/SEDE 1995)

	Boom Clay	Opalinus Clay
mineralogy (weight %)		
clay minerals	60	40-80
illite	20-30	18-36
smectite	10-20	-
chlorite	5-20	6-12
kaolinite	20-30	10-20
mixed illite/smectite	5-10	6-12
mixed chlorite/smectite	5-10	-
quartz	20	18
feldspars	5-10	1
carbonates	1-5	5-20
pyrite	1-5	1
organic carbon	1-5	0.7
hydro-mechanical properties		
total porosity (%)	36-40	3-12
hydraulic conductivity (m/s)	$2 \cdot 10^{-12}$	$< 10^{-11}$
Young's modulus (elasticity) (MPa)	200-400	2000-3000
plasticity index IP (%)	32-51	-

3   Rationale behind field tracer tests

3.1   Tracer tests in Boom Clay

Safety studies and sensitivity analyses (PAGIS, PACOMA, EVEREST) have shown that the Boom Clay layer is the most important barrier preventing radionuclide migration to the biosphere in the multi-barrier concept considered in Belgium for high level radioactive waste. Therefore the objectives of the migration studies performed at SCK·CEN are:

- to understand the basic phenomena governing the mobility of radionuclides;
- to determine their migration parameters;
- to develop the models needed in performance assessment studies to extrapolate the transport of radionuclides over geological time and spacial scales;
- to demonstrate the predictability of radionuclide migration in the Boom Clay and to assess the reliability of these predictions;
- to enhance public acceptance.

During many years of performing experiments, an understanding of basic transport mechanisms in Boom Clay has been developed. Therefore the field tracer tests performed in the HADES underground research facility at Mol concern the two last of the above objectives. The main aim of the field tracer tests is to demonstrate, that on the basis of parameters derived from small-scale (a few cm) laboratory diffusion experiments, we can predict the migration of a tracer injected into the Boom Clay in the HADES facility over a metric scale and a time scale of several years. Thus, the aim of the field tracer tests is not to determine parameters or to derive migration mechanisms, but to validate in situ our knowledge gained from laboratory migration experiments and in situ hydraulic tests.

Laboratory permeability measurements together with small and large scale in situ hydraulic tests have shown the very low hydraulic conductivity of the Boom Clay and the absence of water conducting fractures at the Mol site. Therefore, migration is mainly controlled by molecular diffusion and advection plays only a secondary role. The results of laboratory migration experiments and of sensitivity analyses show that hR (the product of the diffusion accessible porosity and the retardation factor) and the apparent diffusion constant D_a are the key parameters.

3.2   Tracer tests in Opalinus Clay

In contrast to the Boom Clay, which shows no fractures of importance and lithological variations of only minor importance with regard to radionuclide transport, the Opalinus Clay contains joints and faults and shows more important lithological variations. While for the case of radioactive waste disposal in the Boom Clay several performance assessment (PA) studies (PAGIS, PACOMA, EVEREST) have already been performed, for Opalinus Clay no formal complete PA study has yet been performed. This situation results in a different rationale behind the planned first field tracer test in Opalinus clay compared to the running and planned tests in the Boom Clay.

A conceptual model of groundwater flow and solute transport, mainly based on observations in open clay pits, was developed for preliminary safety assessment calculations (Nagra 1988). This model assumes advective-dispersive transport in joints and faults, accompanied by matrix diffusion into the adjacent undisturbed rock. However, hydraulic tests in shallow and deep boreholes indicate a drastic decrease in the hydraulic conductivity of the Opalinus Clay with depth, leading to the question whether joints and faults represent preferential pathways for solute migration also at repository depth (several 100 metres) or if they can be neglected (i.e. diffusion would be the only transport mechanism). Other open questions include the specific role of lithological inhomogeneities (silty, sandy or carbonaceous layers), microfractures and the damage zone around fractures in solute transport.

Given these fundamental open questions regarding Opalinus Clay, initial experiments must concentrate on the earlier objectives in the list given above. Therefore the understanding of groundwater flow and solute transport mechanisms in a highly consolidated fractured claystone formation and the evaluation of parameters for radionuclide transport models are the main objectives of the field tracer experiment planned to be carried out at Mt. Terri.

4   Design of field tracer tests in Boom Clay

The design of the field tracer tests in the Boom Clay is strongly determined by the scale of metres to be studied: tracer tests with retarded species (even moderately retarded) are impossible over this spatial scale as the test would take centuries or even many thousands of years. Therefore, the first field tracer test, the so-called CP1 test, was performed using tritiated water. The test should demonstrate that advection is only of minor importance in the movement of water through the Boom Clay. The experimental set-up consists of a multi-filter piezometer nest containing nine cylindrical filters with a centre-to-centre distance of 1 m (see Fig.1) . The test was installed horizontally through the concrete plug at the end of the gallery. There is a strong hydraulic gradient, and thus flow, towards the gallery caused by its excavation and the permeability of the concrete plug. The tritiated water was injected into the central filter; this should allow the small asymmetry in the diffusion cloud due to the advective flow to be seen. In the undisturbed clay formation the hydraulic gradient is more than a thousand times smaller.

Two field tracer tests of the same type have been installed for the injection of I-125 as I^-. Iodine was chosen as tracer because it is not chemically retarded; as a negative ion is subjected to the effect of anion exclusion (the clay particles are negatively charged) and it thus has a lower diffusion accessible porosity than water. Moreover, performance assessment studies have shown that I-129 is one of the most critical radionuclides. Two piezometers TD41H and TD41V of a similar type to the CP1 piezometer were installed (see Fig. 2) . The laboratory experiments have shown that both the apparent diffusion and the hydraulic conductivity show an anisotropy. This anisotropy is related to the orientation of the clay particles with the bedding plane. In the horizontal plane, i.e. parallel to the bedding, both the hydraulic conductivity and apparent diffusion constant are a factor of two higher than in the vertical direction. Therefore, one piezometer was installed horizontally and one vertically. The vertical piezometer was also used to study another clay horizon. The filters neighbouring the injection filter were placed at a distance of 35 cm because I-125 has a half-life of about 60 days and thus its migration can only be followed for about three years.

A new field tracer test was started at the end of 1995. It is a large-scale 3-dimensional injection experiment with tritiated water and carbon-14-labelled bicarbonate injected as a cocktail. This experiment is called the TRIBICARB-3D test. The test set-up consists of three parallel horizontal piezometers: the injection piezometer, a detection piezometer on the right and one below the injection piezometer (see Fig. 3).The distance between the piezometers at the level of the injection filter is 0.9 and 1.5 metres respectively. The spacing between the fiters on each piezometer is one metre. The reasons for this design are: to follow the contaminant cloud on a scale of metres in the three dimensions and to show that there is no preferential migration pathway along the injection piezometer. Due to the anisotropy in the apparent diffusion constant and the hydraulic conductivity, a contaminant cloud with an ellipsoid shape is expected. The aim of tritiated water injection is the same as for the CP1 experiment. C-14-labelled bicarbonate is injected because performance assessment studies have shown that the potential dose due to C-14 is very sensitive to its retardation factor. The release time for bicarbonate is a few tens of thousands of years. As the half-life of C-14 is only 5730 years, a very small retardation, e.g. a factor of three, is sufficient to cause a large decrease in C-14 release due to radioactive decay.

Further tracer tests are also foreseen in the TD41 piezometers. Within the framework of the EC-NIRAS/ONDRAF project TRANCOM-CLAY it is planned to inject C-14 labelled organic molecules previously extracted from the Boom Clay. The aim of this tracer injection is to follow in situ the migration of naturally present organic molecules which potentially play a role in radionuclide transport. This test will also be supported by laboratory experiments.

5 Preliminary design of field tracer tests in Opalinus Clay

To answer the open questions raised in section 3.2 the following concept for a simple the tracer test has been developed (cf. Fig. 4): a cocktail of appropriate tracers will be injected into a packed-off small-diameter borehole over a long period (2 years or more). The resulting tracer distribution will be mapped or visualised in large-diameter overcored drillcores or in cores from parallel boreholes. The in situ experiment will be started after a tracer evaluation study including laboratory experiments with Opalinus Clay samples. Laboratory experiments are being performed by the Centre d'Energie Nucléaire in Grenoble, by the University of Bern and by the British Geological Survey. Possible tracer candidates under discussion are non-sorbing tracers (I^-, Br^-, EDTA, fluorescent agents), weakly sorbing tracers (Li^+, Mg^{++}, Sr^{++}), stable isotopes, an oxidising tracer ($KMnO_4$) and KCl brine (for resistivity imaging). There is no permit for the application of radioactive tracers at this test site. It is hoped that this tracer experiment will provide data to enable identification of the dominant physical processes (advection-dispersion, diffusion and related conceptual models) as well as information on a range of input parameters used for transport models.

6 Main results and application of tracer tests performed in the Boom Clay

For the CP1, TD41H and TD41V tests in the HADES facility, the progress of the tracer cloud was calculated prior to the start of the test based on the laboratory data. The concentrations in the filters of the CP1 test were predicted for a hundred years and for TD41V and TD41H for three years. The TD41 tests are finished while CP1has now been in progress for more than eight years. The predictions and the experimental results are shown on Fig. 5. The results of the CP1 test correspond well with the predicted values. The results for the TD41 tests also correspond rather well with the predictions. There is however a small shift between the prediction and experimental results for the injection filters. This shift is due to difficulties in the sampling of those filters.

The field tracer tests performed up till now in the HADES facility in the Boom Clay have been very successful and fulfilled the predetermined objectives. With these tracer tests we have been able to demonstrate that diffusion is the dominant transport mechanism, which is the cornerstone of our performance assessment; we have also been able to show in situ, on a scale of metres, the validity of our laboratory data for tritiated water and iodine.

The conceptual model and data used for the prediction of the tracer tests have been used in performance assessment studies e.g. EVEREST (Marivoet et al. 1996). Once the results from the C-14 tests are available they will also be incorporated into ongoing PA studies. The CP1 experiment has been used by five research groups as one of the test cases in the INTRAVAL phase

2 benchmark exercise (NEA, SKI 1996). The good results of the field tracer tests have certainly strenghtened the confidence of both our experimental and PA teams in the migration model and have a positive influence on the acceptability of clay as option for high level waste disposal within the scientific community. The influence of these experiments on public acceptance is however, difficult to assess at this stage.

7   Conclusions

Field tracer tests in clays can provide valuable data at   different stages in a geological waste disposal research project:

- in the initial stage when geological media such as clay are being characterised, field tracer tests can help to build an understanding of the dominant transport mechanisms operating on different scales;

- in a later stage when formal performance assessment studies are carried out, field tracer tests can help to build confidence in the applied migration models and data;

- in particular, field tracer tests can increase confidence that predictions based on performance assessment models and data measured in the laboratory can be justifiably used in analyses of long-term performance;

- it must, of course, be recognised that temporal and spatial extrapolations will always be needed since migration tests must be scaled down to produce measurable results;

- nevertheless, when a project is being defended before the scientific community and general public, successful field tracer tests can increase the project's acceptability.

To perform field tracer tests that fulfil the above objectives, a well intergrated combination of laboratory experiments, site characterisation and model development is a necessity. The tests performed in the HADES facility at Mol show that this is possible and the tests in the Mt Terri Project are planned to cover the first step of this methodology, tackling some basis understanding of flow and transport mechanisms using laboratory experiments and a single-hole field tracer test.

Field tracer tests in clays where diffusion is the dominant transport process have the important limitation mentioned above: it is impossible to follow the migration of moderately or strongly retarded species over a scale of metres in a reasonable time period. For such species the only alternatives are the study of natural analogues or the study of the distribution of e.g. rare earth elements at the site itself to derive their migration behaviour.

References:

Gautschi, A. (1996): Hydrogeology of a Fractured Shale (Opalinus Clay): Implications for radionuclide migration.- In: Proceedings of a joint OECD-NEA/EC workshop on 'Fluid flow through faults and fractures in argillaceous formations', Bern, Switzerland, 10-12 June 1996 (in print).

Gautschi, A., Ross, C. & Scholtis, A. (1993): Porewater - groundwater relationships in Jurassic shales and limestones of northern Switzerland. - In: Manning, D.A.C., Hall, P.L. & Hughes, C.R. (Eds.): Geochemistry of Clay-Pore Fluid Interactions. Mineral. Soc. series 4, 412 - 422. Chapman & Hall, London, ISBN 0-412-48980-5.

Geotechnical Institute Ltd. (1995): International Research Project in The Mt. Terri Reconnaissance Tunnel for the Hydrogeological, Geochemical and Geotechnical Characterisation of an

Argillaceous Formation (Opalinus Clay - Project Proposal. - Nagra Internal Report, Nagra, Wettingen, Switzerland.

Marivoet, J., Volckaert, G., Wemeare, I. & Wibin, J. (1996): Evaluation of Elements Responsible for the Effective Engaged Dose Rates Associated with the Final Storage of Radioactive Waste: EVEREST project. Volume 2a: Clay formation, site in Belgium. European Commission, nuclear science and technology, Luxembourg, EUR 17449 EN

Nagra (1988): Sediment Study - Interim Report 1988: Disposal Options for Long-Lived Radioactive Waste in Swiss Sedimentary Formations (Executive Summary). - Nagra Technical Report NTB 88-25E, Nagra, Baden, Switzerland.

NEA/SEDE (1995): A Catalogue of the Characteristics of Argillaceous Rocks Studied with Respect to Radioactive Waste Disposal Issues. - Unpublished Internal Document of the OECD Nuclear Energy Agency, Paris.

NEA, SKI (1996): The international INTRAVAL project, Developing groundwater flow and transport models for radioactive waste disposal, Six years of experience from the INTRAVAL project, Final results. OECD Nuclear Energy Agency, Paris.

Vandenberghe, N. & Laga, P. (1986): The septaria of the Boom Clay (Rupelian) in its type area in Belgium. Aardkundige Mededelingen, vol. 3.

Van Echelpoel, E. & Weedon, G.P. (1990): Milankovitch cyclicity and the Boom Clay Formation: an Oligocene siliciclastic shelf sequence in Belgium. Geol. Mag. 127 (6).

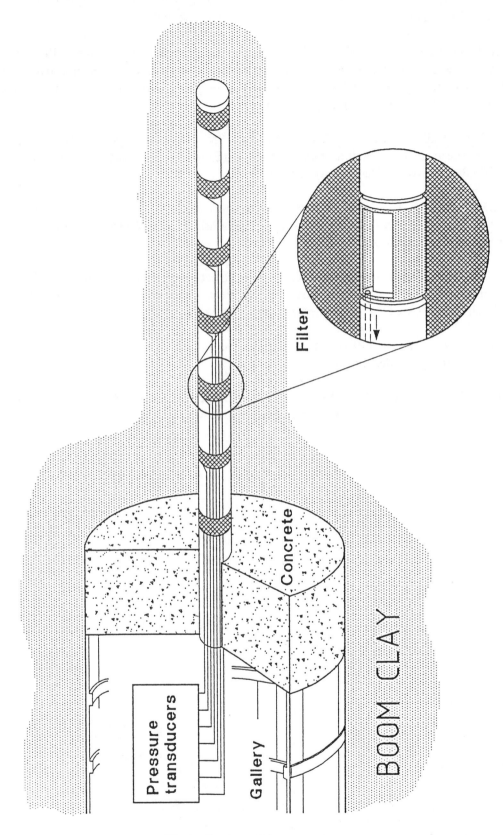

Fig 1: Design of the CP1 test

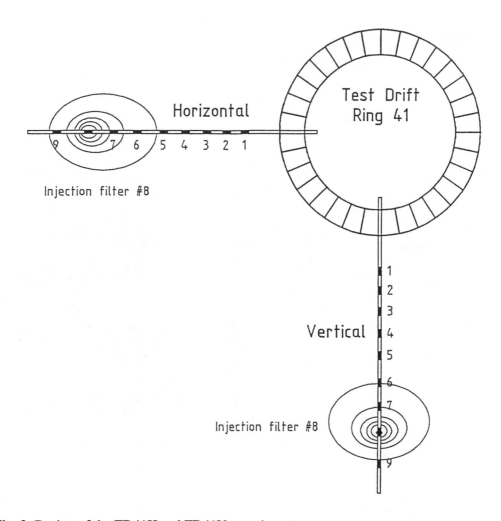

Fig. 2: Design of the TD41H and TD41V experiments

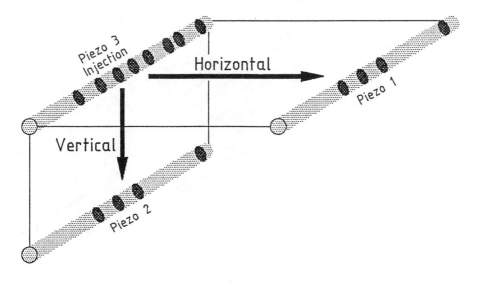

Fig. 3: Design of the TRIBICARB-3D experiment

Mont Terri Project

Groundwater Flow Mechanism

Experiment FM-C: Tracer Injection

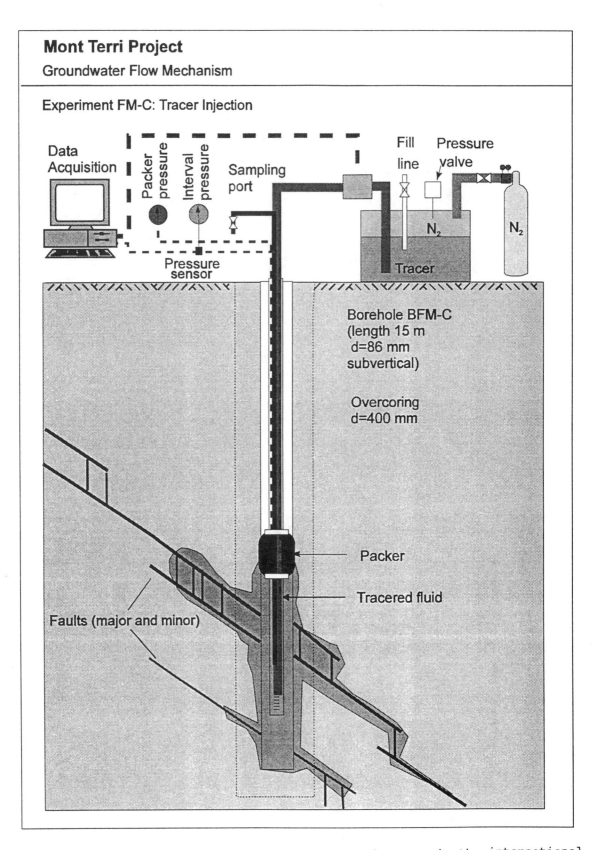

Fig. 4: Configuration of the tracer experiment foreseen in the international Mt. Terri Project

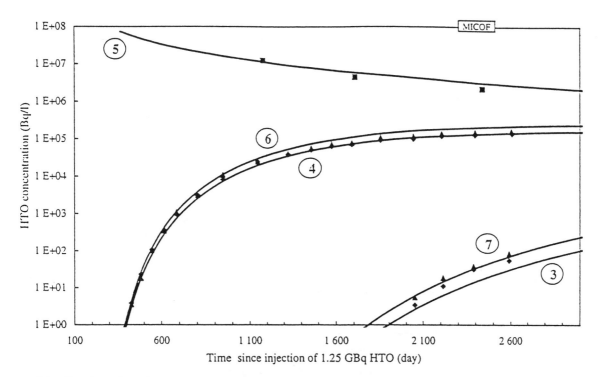

Fig. 5a: Results of the CP1 experiment: HTO concentrations in the interstitial water sampled from filters CP1/3 to CP1/7 (symbols) and MICOF simulations (lines). Linear time scale

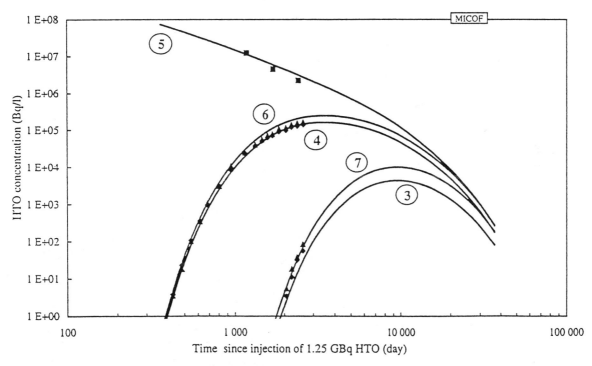

Fig. 5b: Results of the CP1 experiment: HTO concentrations in the interstitial water sampled from filters CP1/3 to CP1/7 (symbols) and MICOF simulations (line). Logarithmic time scale

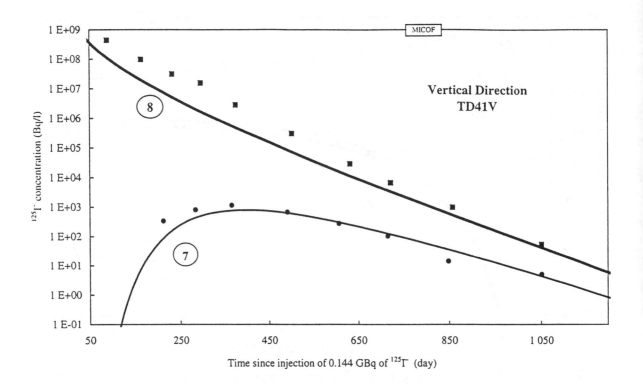

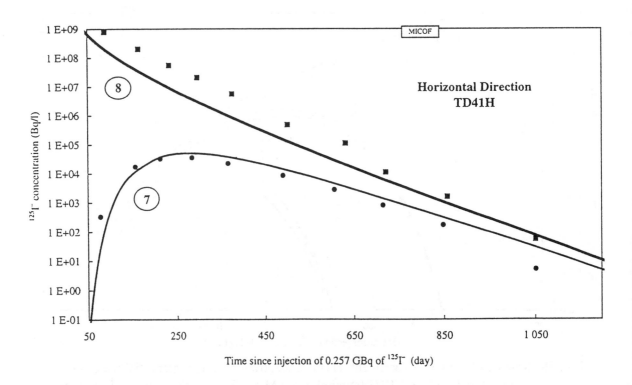

Fig. 6: Results of the TD41 tests: evolution of the measured $^{125}I^-$ concentrations (symbols) and the concentrations predicted with the MICOF code

Rationale for the H-19 and H-11 Tracer Tests at the WIPP Site

Richard L. Beauheim, Lucy C. Meigs, and Peter B. Davies
Sandia National Laboratories, USA

Abstract

The Waste Isolation Pilot Plant (WIPP) is a repository for transuranic wastes constructed in bedded Permian-age halite in the Delaware Basin, a sedimentary basin in southeastern New Mexico, USA. A drilling scenario has been identified during performance assessment (PA) that could lead to the release of radionuclides to the Culebra Dolomite Member of the Rustler Formation, the most transmissive water-saturated unit above the repository horizon. Were this to occur, the radionuclides would need to be largely contained within the Culebra (or neighboring strata) within the WIPP-site boundary through the period lasting for 10,000 years after repository closure for WIPP to remain in compliance with applicable regulations on allowable releases. Thus, processes affecting transport of radionuclides within the Culebra are of importance to PA.

The Culebra is an approximately 7-m-thick, variably fractured dolomite with massive and vuggy layers. Hydraulic tests conducted at 41 well locations showed that the transmissivity of the Culebra varies over six orders of magnitude in the vicinity of the WIPP site. Convergent-flow tracer tests were conducted in the Culebra at three high-transmissivity locations between 1981 and 1988. The breakthrough curves were successfully simulated using a homogeneous double-porosity model with directional anisotropy in permeability. For preliminary performance assessments, models of radionuclide transport treated the Culebra as a double-porosity medium with uniformly spaced horizontal fractures over its entire thickness and spatially varying transmissivity. Simulations with Monte Carlo sampling from varying representations of the Culebra transmissivity (T) field and varying values of fracture spacing, matrix porosity, fracture porosity, matrix distribution coefficients (K_ds), fracture-surface K_ds, radionuclide solubilities, and 42 other parameters were used to define complementary cumulative distribution functions (CCDFs) displaying the probability that radionuclide releases will exceed calculated quantities, as required by regulation. Based on sensitivity studies carried out in 1992, fracture spacing, matrix K_ds, and radionuclide solubilities were found to be "very important" parameters, the T field and matrix porosity to be "important" parameters, and fracture porosity and fracture-surface K_ds to be "less important" parameters in terms of reductions in their uncertainties being able to affect the position of the mean CCDF near the regulatory limits.

The model and parameter ranges used by PA were scrutinized by different review groups, who concluded that there was inadequate experimental justification to rule out alternative models and parameters. Accordingly, a new series of tracer tests was planned to address the review groups' criticisms and provide a defensible model and parameters for PA modelling. A new seven-well location, the H-19 hydropad, was established for the testing. The test design focused on improving our understanding of matrix diffusion, the effects of layering in the Culebra on transport, and the causes of directional differences in transport. The tracer tests were linked to a field hydraulic-testing program and laboratory studies measuring radionuclide solubilities, sorption on the Culebra matrix, and matrix porosity, tortuosity, and permeability to address all of the "very important" and "important" transport-related parameters identified from the 1992 PA sensitivity analyses. The data and, in some instances, revised conceptual models resulting from these studies were used in the performance assessment that was a part of the formal certification application for the WIPP submitted by the U.S. Department of Energy to the U.S. Environmental Protection Agency in October 1996.

Introduction

The Waste Isolation Pilot Plant (WIPP) is a repository proposed by the U.S. Department of Energy (DOE) for disposal of defense-related transuranic wastes. The WIPP has been constructed in bedded Permian-age halite in the Delaware Basin, a sedimentary basin in southeastern New Mexico, USA (Figure 1). Because the Delaware Basin is the target of active drilling for energy resources, performance assessment (PA) for the WIPP must consider the possibility of inadvertent human intrusion at a time in the future when the existence of the WIPP is assumed to have been forgotten. Based on the understanding and knowledge of WIPP-site hydrogeology developed during site characterization, a drilling scenario has been identified that could lead to the release of radionuclides to the Culebra Dolomite Member of the Rustler Formation, the most transmissive water-saturated unit above the repository horizon (Figure 2). In this scenario, one borehole penetrates the repository, located 655 m below ground surface in bedded halite of the Salado Formation, and continues into the underlying Castile Formation, where it encounters a brine reservoir. Brine reservoirs have been encountered in approximately 30 of the 350+ boreholes drilled through the evaporite section in the Delaware Basin in the vicinity of the WIPP site. The reservoirs are pressurized and flow brine at the ground surface at initial rates between 100 and 3000 m^3/day [1]. The presence of one of these brine reservoirs beneath the WIPP repository is considered unlikely, but cannot be ruled out.

In some of the intrusion scenarios considered in the PA, the borehole penetrating the brine reservoir is plugged above the repository, leaving the repository connected to the brine reservoir. A second borehole is then hypothesized to penetrate the repository at another location. At some time in the future, whatever materials were used to plug the second borehole are assumed to degrade, and a flow path is established from the brine reservoir through the repository and into the Culebra. Were this to occur, any dissolved radionuclides that were transported up the borehole into the Culebra would need to be largely contained within the Culebra (or neighboring strata) within the WIPP-site boundary through the period lasting for 10,000 years after repository closure for WIPP to remain in compliance with applicable regulations on allowable releases. Thus, processes affecting transport of radionuclides within the Culebra are of importance to the WIPP project.

Figure 1. Location of the WIPP site

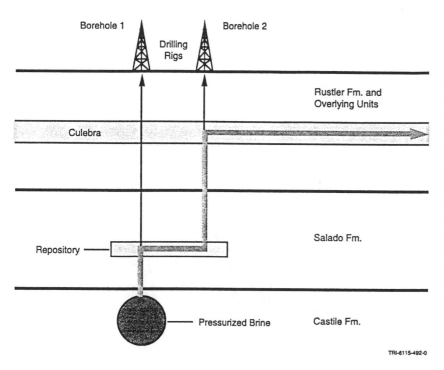

Figure 2. **Drilling scenario for radionuclide release from the WIPP**

The Culebra is an approximately 7-m-thick, variably fractured dolomite with massive and vuggy layers. During site characterization from 1974 through 1988, 61 wells were completed to the Culebra at 41 locations around the WIPP site (Figure 3). Hydraulic testing was performed at each of these locations to determine the local transmissivity of the Culebra and to provide information on the regional distribution of transmissivity. At some of these locations (see Figure 3), open fractures were observed in Culebra core and hydraulic tests indicated relatively high transmissivity. Hydraulic tests at these locations could be readily interpreted using double-porosity models that assumed that most of the permeability of the medium was provided by fractures while most of the storage capacity was provided by the primary porosity of the rock matrix between the fractures. At other locations (see Figure 3), the fractures in the Culebra were found to be sealed with gypsum and the Culebra had relatively low transmissivity. Hydraulic tests at these locations were most simply interpreted using single-porosity models for flow through a porous medium.

The hydraulic tests showed that the transmissivity of the Culebra varies over at least six orders of magnitude in the vicinity of the WIPP site. To represent this heterogeneity in flow and transport simulations, multiple realizations of the transmissivity (T) field are generated. Initial information for the T fields comes from the transmissivities and steady-state hydraulic heads measured at the individual wells. The T fields are calibrated by comparing simulated responses to the transient pressure/water-level responses observed while performing large-scale pumping tests of several months' duration and while sinking the shafts at the WIPP site. An automated inverse code, GRASP-INV [2, 3] uses pilot points (synthetic measured-transmissivity locations) to improve the model fit to the observed data until acceptance criteria are met. Using GRASP-INV, as many different but equally likely representations of the T field can be generated as desired (Figure 4).

Convergent-flow tracer tests were conducted in the Culebra at three of the high-transmissivity (double-porosity) locations between 1981 and 1988. The tracer-breakthrough curves obtained from these tests showed two major characteristics, as exemplified by the data from the H-6 test (Figure 5): First, the rate of transport was directionally dependent, with rapid transport and relatively high peak concentrations occurring along one direction while slower transport and lower peak concentrations occurred along another direction. Second, the breakthrough curves showed long "tails", believed to reflect the influence of some type of physical-retardation mechanism, most likely matrix diffusion. The breakthrough curves were successfully simulated (Figure 6) using a homogeneous double-porosity model with uniform cubic matrix blocks and directional anisotropy in permeability [4]. The geometric conceptualization of the fracture-

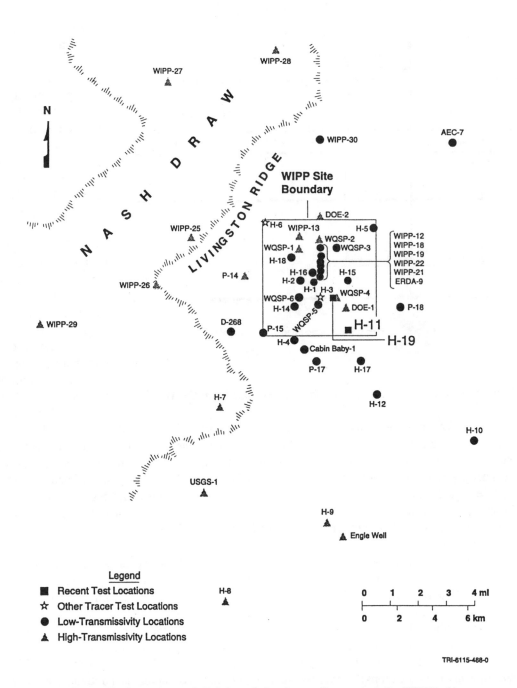

Figure 3. Locations of Culebra dolomite wells around the WIPP site

matrix system used in the model was found to be unimportant, as long as the ratio of fracture surface area to matrix volume was held constant and diffusion did not reach the center of the matrix blocks during the simulation. Thus, cubic matrix blocks (three orthogonal fracture sets) could be replaced with slabs (a single set of horizontal fractures) by decreasing the fracture spacing by a factor of three.

Preliminary Performance Assessment

For preliminary performance assessments, models of radionuclide transport treated the Culebra as a double-porosity medium with uniformly spaced horizontal fractures over its entire thickness and spatially varying transmissivity. The 1992 preliminary PA performed 70 simulations of the entire

110

repository/geosphere system with Monte Carlo sampling of varying representations of the Culebra transmissivity field and varying values of fracture spacing (matrix block length), matrix porosity, fracture porosity, matrix distribution coefficients (K_ds), fracture-surface K_ds, radionuclide solubilities, and 42 other parameters. For each simulation, a complementary cumulative distribution function (CCDF) was calculated displaying the probability that radionuclide releases will exceed calculated quantities, as required by regulation (Figure 7). Based on sensitivity studies [5], the various sampled parameters were ranked as "critically important", "very important", "important", and "less important" in terms of their

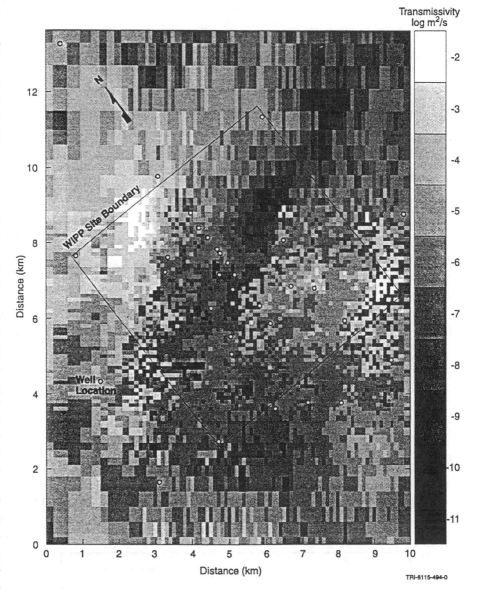

Figure 4. Transient-calibrated transmissivity field no. 77

potential to affect releases of radionuclides from the site. For parameters in the first three categories, reductions in uncertainties have the potential to shift the position of the mean CCDF near the regulatory limit. The sensitivity analyses showed fracture spacing, matrix K_ds, and radionuclide solubilities to be "very important" parameters, the transmissivity field and matrix porosity to be "important" parameters, and fracture porosity and fracture-surface K_ds to be "less important" parameters.

Two questions arose from the PA sensitivity studies: (1) Is the interpretive model applied to the tracer-test data, and extended to PA, defensible? (2) If so, can the ranges of parameter values used by PA be justified and defensibly narrowed if necessary and, in particular, can we confidently quantify the amount of diffusion we expect to occur? To gain insight into these questions, the tracer-test interpretations and PA modelling were presented to numerous review and regulatory groups, including INTRAVAL, the U.S. National Academy of Sciences, the U.S. Environmental Protection Agency (EPA), and the New Mexico Environmental Evaluation Group. Comments received from these groups indicated that the data were not adequate to rule out alternative conceptual models for transport that might have different

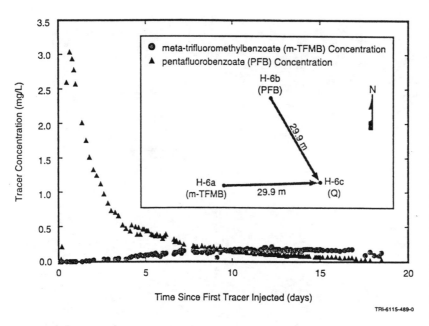

Figure 5. Breakthrough curves for the H-6 convergent-flow tracer test

implications when applied to PA. Specific comments pertained to: (1) the interpretive model's reliance on matrix diffusion as the sole cause of the tailing of the breakthrough curves, neglecting the possible effects of complexities in the tracer source term, small-scale heterogeneity in permeability along the travel paths, and diffusion into stagnant (or slow-moving) water within low-aperture or poorly connected parts of fractures; (2) the validity of the assumption that the Culebra is vertically homogeneous; and (3) reliance on hydraulic anisotropy to explain directional differences in transport.

New Tracer Tests

As a result of these comments, an expanded field tracer-testing program was developed [6], with strong ties to laboratory programs. The scope of the field program was developed through extensive interactions with the review groups and other interested scientists. The primary objectives of the tracer-testing program were to: (1) test for the occurrence of matrix diffusion in the Culebra; (2) quantify or bound the amount of matrix diffusion occurring, which involves a combination of fracture spacing, matrix porosity, and matrix tortuosity; (3) evaluate the effects of layering within the Culebra on flow and transport; and (4) investigate the causes of directional differences in transport within the Culebra. We anticipated that by meeting these objectives, we would be able to either verify that the model previously used was adequate or develop a new model, as well as develop defensible ranges for the parameters needed for modelling.

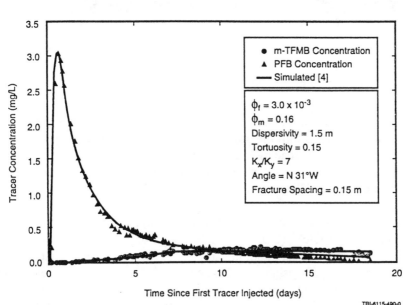

Figure 6. Anisotropic double-porosity simulation of H-6 breakthrough curves

Four elements in the design of the new tests were focused on the demonstration and quantification of

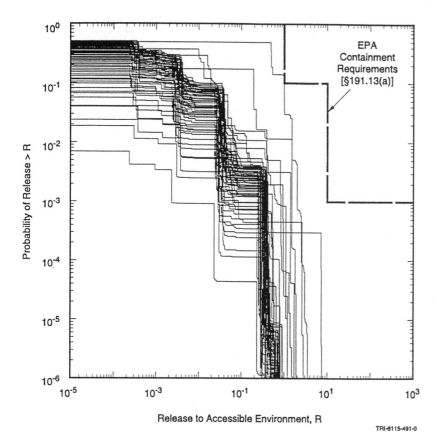

Figure 7. Example of CCDFs for normalized radionuclide release to the accessible environment from the WIPP [5]

matrix diffusion. First, a single-well injection-withdrawal (SWIW) tracer test was planned which was expected to distinguish clearly between the effects of matrix diffusion and heterogeneity in permeability. Pretest simulations comparing the effects of matrix diffusion and heterogeneity showed that the recovery curves were significantly different for single-porosity as compared to double-porosity conditions. Double-porosity simulations produce a recovery curve with tracer concentration decreasing asymptotically as $1/t^{3/2}$, which appears as a constant slope of -3/2 on a log-log plot of normalized tracer concentration (C/C_0) vs. time since injection (t) (Figure 8). All single-porosity models, regardless of the degree of heterogeneity, showed more rapid tracer-mass recovery than the models that included matrix diffusion (Figure 9). Second, two different conservative tracers having different free-water diffusion coefficients were to be injected together in a convergent-flow tracer test to show the effects of different amounts of diffusion (Figure 10). Third, tracer injections were to be repeated in convergent-flow tests

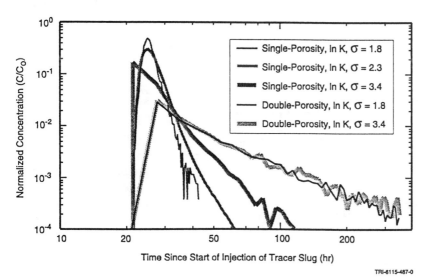

Figure 8. Example tracer-recovery curves for single-well test: heterogeneous, single- and double-porosity models

performed at different pumping rates to show the effects of velocity on diffusion (Figure 10). Fourth, a tool was designed that would allow (1) release of tracer downhole without overpressurizing the wellbore and (2) measurement of the downhole tracer concentration in an injection well as a function of time. By comparing the breakthrough curve obtained using this type of "passive" injection and a known source term to those obtained using pressurized "slug" injections with uncertain source terms, we hoped to

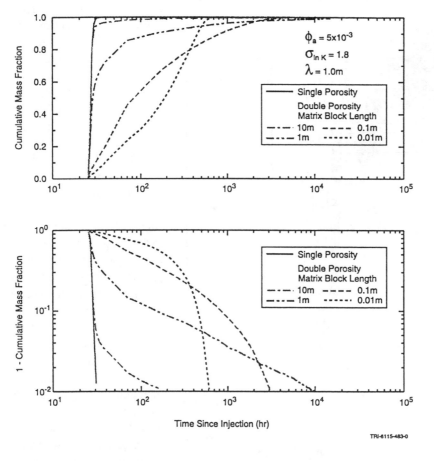

Figure 9. Example mass recovery curves for single-well test using heterogeneous, single- and double-porosity models

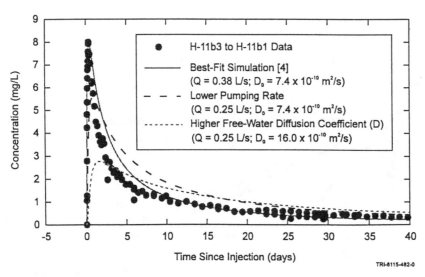

Figure 10. Simulations showing effects of decreased pumping rate and increased diffusion coefficient

determine if we were inappropriately attributing the effects of source-term complexities to matrix diffusion. Unfortunately, problems with this tool prevented its use during the tracer tests.

To evaluate the significance of layering in the Culebra, tracer injections were planned into the upper and lower portions of the Culebra to provide breakthrough curves to compare to those obtained from injections over the full thickness. To evaluate whether directional differences in transport were caused by anisotropy alone or combined with heterogeneity, tracer injections were planned in six wells at different distances and orientations from a central pumping well. If all six tracer-breakthrough curves could not be matched using a single anisotropic model, heterogeneity would be assumed to be a significant factor in transport.

A new seven-well location, the H-19 hydropad, was established for the tracer testing (Figure 11). The H-19 hydropad was located along what is believed to be the likely groundwater pathway in the Culebra from above the repository to the WIPP site boundary. Tracer testing was planned in stages, with a preliminary test to be performed after four wells had been completed to aid in selecting locations for the remaining three wells. The

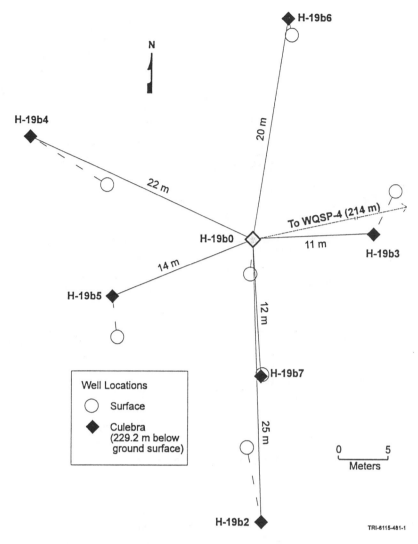

Figure 11. Well locations at the H-19 hydropad

information obtained from the preliminary test would also aid in the design of a more comprehensive series of tests to be performed involving the full seven-well array. Based on the information obtained from the preliminary test, additional testing was also planned for one of the locations previously tested, the H-11 hydropad, to resolve uncertainties associated with the previous test at that location. Information on the results of the tracer testing at the H-19 and H-11 hydropads is presented in a companion paper [7].

Associated Testing

In conjunction with the tracer-testing program, a hydraulic-testing program was established to: (1) characterize the between-well hydraulics of the Culebra at the H-19 hydropad; (2) measure vertical variations in permeability at the H-19 hydropad; and (3) allow for better definition of the Culebra transmissivity field. The information obtained in meeting the first two objectives listed above was used in refining the design of the tracer tests at the H-19 hydropad. The between-well hydraulics at the H-19 hydropad were evaluated by performing cross-hole sinusoidal pumping tests. For these tests, packers were set in each of the wells on the hydropad to divide the Culebra into upper and lower segments. Four tests were then performed by successively pumping the upper and lower Culebra in wells H-19b0 and H-19b4 while monitoring pressure responses in the other isolated zones in the various wells. The pumping began with a constant-rate phase to establish a dominant pressure trend at the hydropad, and then shifted to a sinusoidal rate sequence. The sinusoidal pumping rates produced sinusoidal pressure responses in the observation zones that could be readily distinguished from on-going responses to other hydraulic stresses (Figure 12). The test responses showed the upper and lower Culebra to be poorly connected over at least a portion of the H-19 hydropad. Also, anisotropy alone probably could not explain the directional differences observed in the responses.

Vertical variations in permeability were evaluated using three different techniques. First, water production was monitored during drilling in those wells in which the Culebra was cored using compressed air as the circulation medium, while pressures in nearby wells were monitored to see if the responses differed as different parts of the Culebra were penetrated. Little water production and negligible pressure responses were observed when the upper approximately 3 m of Culebra was drilled. Water production and

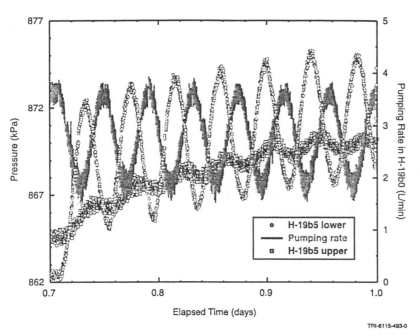

TRI-6115-493-0

Figure 12. Pressure responses in monitoring well H-19b5 and rate sequence in lower zone of active well H-19b0

pressure responses increased significantly as the lower portion of the Culebra was drilled (Figure 13). Second, upper and lower Culebra permeabilities were assessed from the sinusoidal pumping tests discussed above. Third, hydrophysical logging [8] was conducted in three of the H-19 boreholes. This logging confirmed the absence of significant flow in the upper few meters of Culebra and showed unevenly distributed flow throughout the lower section of the Culebra (Figure 14).

Because a convergent-flow tracer test also constitutes *de facto* a long-term pumping test, pressures/water levels were monitored during the tracer test in all observation wells that had responded to previous tests on the H-19 hydropad. The transient responses observed in these wells were used to improve the calibration of the Culebra T fields.

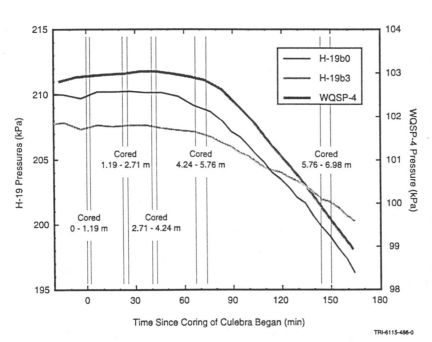

TRI-6115-486-0

Figure 13. Pressure changes in observation wells during coring of H-19b2

The field program was also integrated with laboratory programs measuring radionuclide solubilities, sorption on the Culebra matrix, and matrix porosity, tortuosity, and permeability. The combination of these field and laboratory programs addressed all of the "very important" and "important" transport-related parameters identified from the 1992 PA sensitivity analyses. The data and, in some instances, revised conceptual models resulting from these studies were used in the performance assessment that was a part of the Compliance Certification Application submitted by the DOE to the EPA in October 1996.

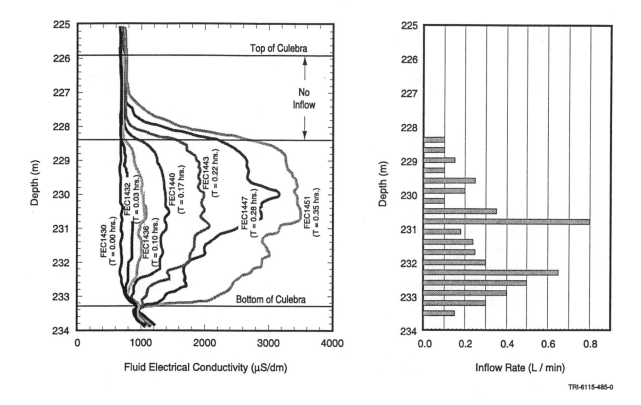

Figure 14. Hydrophysical logging results and inflow simulation for H-19b0

Summary

In summary, information gathered during site characterization was used to develop a conceptual model of the hydrogeology of the WIPP site. Regulatory requirements led to the consideration of a pathway from a Castile brine reservoir through the repository to the Culebra. Preliminary PA modelling of this pathway showed that WIPP's compliance with regulatory release limits could be affected by the model and parameter ranges used to describe transport in the Culebra. The model and parameter ranges used by PA were scrutinized by different review groups, who concluded that there was inadequate experimental justification to rule out alternative models and parameters. Accordingly, a new series of tracer tests was planned to address the review groups' criticisms and provide a defensible model and parameters for PA modelling. The test design focused on improving our understanding of matrix diffusion, the effects of layering in the Culebra on transport, and the causes of directional differences in transport. The tracer tests were linked to other field and laboratory studies providing additional data needed by PA. The data and, in some instances, revised conceptual models resulting from these studies were used in the performance assessment that was a part of the formal certification application for the WIPP submitted by the DOE to the EPA in October 1996.

Acknowledgment:

This work was supported by U.S. DOE under contract number DE-ACO4-94AL85000.

117

References

[1]   Popielak, R.S., Beauheim, R.L., Black, S.R., Coons, W.E., Ellingson, C.T., and Olsen, R.L., Brine Reservoirs in the Castile Formation, Waste Isolation Pilot Plant (WIPP) Project, Southeastern New Mexico, TME 3153, 1983, U.S. Department of Energy.

[2]   RamaRao, B.S., LaVenue, A.M., Marsily, G. de, and Marietta, M.G., Pilot Point Methodology for Automated Calibration of an Ensemble of Conditionally Simulated Transmissivity Fields, 1, Theory and Computational Experiments, Water Resources Research, 1995, 31, 475-493.

[3]   LaVenue, A.M., RamaRao, B.S., Marsily, G. de, and Marietta, M.G., Pilot Point Methodology for Automated Calibration of an Ensemble of Conditionally Simulated Transmissivity Fields, 2, Application, Water Resources Research, 1995, 31, 495-516.

[4]   Jones, T.L., Kelley, V.A., Pickens, J.F., Upton, D.T., Beauheim, R.L., and Davies, P.B., Integration of Interpretation Results of Tracer Tests Performed in the Culebra Dolomite at the Waste Isolation Pilot Plant Site, SAND92-1579, 1992, Sandia National Laboratories.

[5]   WIPP Performance Assessment Department, Preliminary Performance Assessment for the Waste Isolation Pilot Plant, December 1992, Volume 4: Uncertainty and Sensitivity Analyses for 40 CFR 191, Subpart B, SAND92-0700/4, 1993, Sandia National Laboratories.

[6]   Beauheim, R.L., Meigs, L.C., Saulnier, G.J., Jr., and Stensrud, W.A., Culebra Transport Program Test Plan: Tracer Testing of the Culebra Dolomite Member of the Rustler Formation at the H-19 and H-11 Hydropads on the WIPP Site, 1995, Sandia National Laboratories.

[7]   Meigs, L.C., Beauheim, R.L., McCord, J.T., Tsang, Y.W., and Haggerty, R., Design, Modelling, and Current Interpretations of the H-19 and H-11 Tracer Tests at the WIPP Site, 1996, [this volume].

[8]   Pedler, W.H., Evaluation of Interval Specific Flow and Pore Water Hydrochemistry in a High Yield Alluvial Production Well by the HydroPhysical™ Fluid Logging Method, Eos, Transactions, American Geophysical Union, 1993, 74, 304.

The Grimsel Field Tracer Migration Experiment – What Have We Achieved After a Decade of Intensive Work?

W. Russell Alexander
GGWW, University of Berne, Switzerland

Ian G. McKinley, Urs Frick
Nagra, Switzerland

Kunio Ota
PNC, Tono Geoscience Centre, Japan

Executive Summary

Introduction

The Nagra/PNC field tracer migration experiment, carried out in association with PSI (Switzerland) and GSF (Germany), began in 1985 with hydrogeological characterisation of a water conducting shear zone in the granodiorite of Nagra`s Grimsel Test Site (GTS, in the central Swiss Alps) and finished in the spring of 1996. The intervening decade has seen a large series of field tracer migration experiments carried out at the site, increasing in complexity from simple, non-sorbing tracers (uranine, ^{82}Br, ^{123}I, ^{3}He and tritium) through various weakly sorbing tracers (^{22}Na, ^{24}Na, ^{85}Sr and ^{86}Rb) to a final, long-term experiment with strongly sorbing ^{137}Cs. Over the last ten years, the experimental methodology has matured as has our understanding of both the site and the processes influencing *in situ* radionuclide retardation in fractured crystalline rocks. However, this knowledge has been won only following significant investement of both effort and funds so it is appropriate to now review the returns on the investment in terms of a repository performance assessment (PA).

1. General context of the Grimsel Migration Experiment (MI)

The general context in which the experiment should now be judged was the desire to improve confidence in the use of predictive models in a repository PA. Few people, even those involved in the disposal of radioactive waste, fully appreciate the difference between blind testing of model *predictions* and testing if a model can *simulate* particular observations - as can be clearly seen in the literature. This is a crucial point, as pointed out by Pate et al, (1994) "This aspect of blind (ie *predictive*) testing is particularly important as, in many cases, the manner in which the simulation is carried out can be very objective and, if the "answer" is known, can be biased either consciously or subconsciously". In a repository PA, simulation of data brings little or no confidence that the models involved can later predict repository evolution: confidence can be much better built by carrying out a series of predictive modelling exercises followed by experimental runs and a final assessment of the accuracy of the predictions. This coming together of transport modellers, field and laboratory experimenters and (to a lesser extent) performance assessors has been the hallmark of the MI experiment.

2. Specific Objectives

These included testing the applicability of current PA transport codes to quantify radionuclide migration in a real flow system (and, later, the development of a new transport code); the identification of the relevant transport processes for consideration in transport models and assessing how successfully laboratory sorption data may be extrapolated to *in situ* conditions. A final (normally unstated) objective was the indoctrination of staff into the mind-set required for them to make predictions of radionuclide behaviour *in situ* (whether the predictions related to hydrology, transport calculations, geochemistry or flow path geometry) when required.

3. Relation to the overall R+D programme

The MI experiment has been the single biggest experiment to date in the Nagra R+D programme and had a web of connections to other areas of the Swiss and Japanese programmes. As already noted above, there was direct input into Nagra`s laboratory sorption programme where an assessment was made of the relevance of laboratory produced sorption data to *in situ* retardation of radionuclides. In addition, there was much cross-fertilisation between the MI experiment and site characterisation/PA in the field of flow path description.

In the PR field, initial use of the MI experiment was limited but this changed with the production of a Nagra Bulletin on the GTS which included an extensive article on the MI experiment (Frick et al, 1988). Some 35,000 people have now visited the GTS and the MI experiment is a routine stop on the tour of the laboratory. Currently, Nagra is producing a video on the GTS which will include footage on the MI experiment and its successor the Radionuclide Retardation Project (or RRP). Finally, the Federation of Electric Power Companies of Japan recently shot footage for inclusion in a PR video about underground rock laboratories worldwide. Arguably, more remains to be done.

4. Uses and extrapolation of the so obtained information

The most important use of the MI experiment has been the development of testing methodologies and the application of those methods to confidence building within PA. Indeed, the recent Kristallin-1 safety assessment (Nagra, 1994) carried out by Nagra specifically mentions the contribution of the MI experiment to model testing in general. Further, it was noted that "...the results provide confidence in the dual-porosity concept as an appropriate foundation for a model of transport in fractured porous media". In addition, it was noted that "...the model provides a satisfactory interpretation of the measured data and no evidence has been found which would indicate that processes relevant to safety assessment and not accounted for in the model are operating ."

Some effort has gone into extrapolating data on retardation mechanisms from MI to repository relevant host rocks, but this has been limited in some instances. For example, it was noted in Kristallin-1 that, "The key mechanism of matrix diffusion has, in many experiments, been identified as important and its existance and effectiveness are much better founded than 10 years ago.". Despite this, the diffusion constants for the rock matrix used in Kristallin-1 were "...selected on the basis of a survey of (*laboratory*) experimentally determined diffusion constants for crystalline rocks." (authors' italics). Further, no reference is made to evidence from the MI experiment when depths of accessible wall rock are considered other than in the case of one parameter variation where data from MI supporting experiments are used to define a minimum depth of diffusion.

While the work on investigating the connection between laboratory measured sorption data and field retardation has shown that, with enough background information on the flow field, it is possible to show reasonable agreement within the MI experiment between field and laboratory data, this has been taken no further as yet and has most certainly not been utilised in recent Nagra or PNC assessments.

5. Greatest sucesses and failures

Apart from the comments noted above, the greatest success has been the rigorous testing of the PSI developed PA transport code RANCHMD (Hadermann and Roesel, 1985). Of particular note have been the attempts to minimise the number of free parameters in the code by including as many hard data on the shear zone structure, flow paths, tracer sorption values etc as possible (see Heer and Hadermann, 1994, for details). In this way, it has been possible to identify the in situ retardation mechanisms in a more thorough manner than has previously been the case.

The greatest failure has been in the rigorous testing of the PA transport code RANCHMD. The problem here is not the code, rather the limitations imposed on the code by the experiment. PA transport codes such as RANCHMD are specifically developed to calculate the long-term, slow movement of radionuclides in the groundwaters of a repository site. Unfortunately, outwith experiments such as the RRP (see Alexander et al., 1996a,b), field tracer migration experiments cannot provide analogous conditions against which to test such transport codes. In the case of MI, for example, most experiments lasted days to weeks and even the longest experiment conducted, the last ^{137}Cs migration, no more than 20 months or so. This means that testing matrix diffusion within a code such as RANCHMD is relegated to observations based on highly porous matrix (or fracture fill) which may not be of much relevance to a given repository host rock. Also, kinetic effects may play an important role in the experiment but, obviously, are of no relevance to a repository PA.

6. Assessing the results in the light of the original rationale

As noted in section 1 above, the original rationale was to improve confidence in predictive modelling as applied to a repository PA. To this end, the MI experiment has been a great success for Nagra, PNC and the various contractors involved in MI (and, perhaps to a lesser extent, PA) in that it has built a culture of rigorous model testing and, perhaps more importantly, predictive modelling. Of course, precise measurement of such a change in attitude within a disposal programme is difficult but success can always be judged on the quality of the literature on model testing from before and after an experiment such as MI.

7. Potential changes and improvements in design

To look at something in hindsight is always an easy way to build an experimental programme and in MI several things would almost certainly be changed (for example, more complete hydrological characterisation of the experimental site, an earlier structural and petrological description of the flow path). However, a more realistic question might be "knowing now as little as you knew ten years ago, at the beginning of MI, would you do it differently?". In this case, it is likely that we would change much less: the entire experiment has been a learning experience for most of the people involved and has certainly contributed to our views on blind predictive testing and the development and testing of conceptual models of groundwater flow. One weakness, which has perhaps only now been acknowledged, is that, while the field experimenters, laboratory experimenters and transport modellers were in it together from the very beginning, the performance assessors were remarkable only by their absence. This would probably be the single greatest improvement possible to ensure the production of PA relevant data from any field tracer experiment - and the eventual inclusion of such data in a repository PA.

Acknowledgements

Thanks to our numerous colleagues in the MI and RRP projects who have contributed to the success of the work over the last decade. Thanks also to Nagra and PNC for funding the projects.

References

Alexander et al., (1996a) GEOTRAP-FTTE Report, NEA OECD, Paris, France.

Alexander et al., (1996b) Grimsel Test Site 1996, Nagra Bulletin No 27, Nagra, Wettingen, Switzerland.

Frick et al., (1988) Nagra Bulletin, Special Edition 1988, Nagra, Wettingen, Switzerland.

Hadermann and Roesel, (1985) EIR Bericht Nr. 551, PSI, Würenlingen, Switzerland.

Heer and Hadermann, (1994) PSI Bericht 94-13, PSI, Würenlingen, Switzerland.

Nagra, (1994) Kristellin-1, Nagra Technical Report Series, NTB 93-22, Nagra, Wettingen, Switzerland.

Pate et al., (1994) CEC Report EUR 15175EN, CEC, Brussels, Belgium.

Tracer Tests for Site Characterisation?

Juhani Vira and Timo Äikäs
Posiva Oy, Finland

1. INTRODUCTION

Following a country-wide screening and regional assessment a geologic site characterisation programme for final repository of spent nuclear fuel was launched in 1987 at five sites in various part of Finland. The preliminary programme phase was completed and reported in 1992 with the conclusion that a safe repository could be built at any of the five sites studied and there were no significant differences between the sites with bearing on long-term safety but with respect to the efficiency of future more detailed characterisation three of the sites could be preferred to the others. On this basis, since 1993 the studies have been confined to three areas: Olkiluoto in Eurajoki municipality, Kivetty in Äänekoski and Romuvaara in Kuhmo.

The programme for more detailed characterisation includes extensive hydrogeochemical and hydrogeological investigations at the sites. A site-scale structural modelling combined with borehole measurement programme with local water conductivity measurements and pumping and interference tests were used as a basis for the large-scale flow simulations performed as a part of the preliminary site characterisation phase. In the detailed investigation phase a considerable amount of new data has been collected from new boreholes; in addition new measurement methods have been introduced. For instance, the new Posiva Flow Meter has now been used to measure flow rates and, flow directions and water conductivities of the conductive parts of the deep boreholes with 2-m packer intervals. Interference tests have been continued. In parallel with the measurement programme, new updated structural models are being finalised for all the three sites under study.

So far no tracer tests have been performed at these sites. However, the identified connections between two boreholes, some 130 metres apart from each other, would offer a promising site for testing equipment and methods for field scale tracer experiments. At the same time it gives a practical context of judging the rationale and assessing the benefit of tracer tests for site characterisation purposes.

Before considering the rationale for tracer tests it is reasonable to ask what the rationale for site characterisation is, in particular, is site characterisation itself a goal or an instrument? A safety-oriented, instrumental view might consider it as an input for the performance assessment, and in this opinion, the rationale of the tracer tests should be looked at against the needs for performance analysis. On the other hand, it is fairly obvious that a good site description can be considered as a goal itself, since without a consistent overall picture of the site the credibility of any detailed information may be hard to judge. In this paper, therefore, the focus is how tracer tests can facilitate site description and understanding. Nevertheless, the "goodness" of site description, of course, refers to instrumental value, which should also be borne in mind.

Another thing is how to judge the benefits of tracer tests. We might ask whether tracer tests are *useful* for site characterisation – in which case the question is about optimisation and of managerial nature – but we may also ask whether tracer tests are necessary for obtaining some specific piece of information. The latter question is not only managerial but also a scientific one.

2. EXISTING EXPERIENCE

Till today fairly limited experience exists on using tracer tests for site characterisation for waste repositories. Most of the tracer tests performed during the 1980's and the early 1990's are rather limited in spatial extension and the focus has been on transport processes rather than the transport media characterisation.

One of the earlier tracer tests with possible relevance for site characterisation was made in Savannah River 25 years ago with a scale as long as 540 metres [1]. The test appears to have contributed to conceptualisation of the fracture zone studied as a sufficiently homogenous feature for flow modelling. The tracer tests also indicated some problems in earlier hydraulic measurements.

Finnsjön was not intended for a repository site but it was subject to an extensive fracture zone characterisation including tracer tests in the 1980's. Hydraulic testing and indications played obviously a significant role in revealing the perhaps most interesting feature, the horizontal fracture zone 2. Tracer tests were then used for subsequent characterisation of this feature and produced a lot of information on such things as anisotropy, heterogeneity and connectivity of the feature. The tests were also used as cases for extensive modelling effort in the International Intraval Project during several years and in this context they were also used for derivation of different transport parameters.

The Äspö LPT-2 large-scale pumping and tracer test was a major hydraulic and tracer experiment. It generated a lot of information that was subsequently used for testing different structural model assumptions. It confirmed some connections between fracture zones and indicated needs for modifications in the structural model. Nevertheless, the utility of the tracer test performed was limited. Most of the tracers never appeared in the pumping hole during the observation period and what arrived could be interpreted in a thousand of ways.

At the WIPP site in New Mexico a considerable experience exists. The rationale and results of the latest test have been reported in two papers of this Workshop [2], [3]. The conclusion is that the tracer tests have truly contributed to the evolution of the site conceptualisation and the tests can be described at least as useful.

In Finland tracer tests were in the programme of site investigations both at Olkiluoto and at Loviisa, where the two low-and intermediate-level radioactive waste repositories now are situated. The information produced by these tests was of minor value, but helped to understand the nature of the hydraulically interesting features of the sites.

3. USEFUL INFORMATION AND EXPERIENCE WAS OBTAINED BUT...

Most of the field tracer tests have probably produced some results of interest, although sometimes the most significant product may have been the knowledge of how not to do tracer tests. In most cases they have brought out some features of relevance for site understanding. The more difficult question is whether they were crucial for that understanding, and whether the same information could have been obtained with some other methods, possibly by some cheaper or more conclusive methods.

It is obvious that the information obtained with tracer tests in the Finnish site investigations for low- and intermediate waste repository could also have been obtained by hydraulic testing. The tracer test in Olkiluoto revealed the importance of pegmatite veins as potential transport pathways for further studies, but they were in no way essential for understanding the local flow situation.

At Finnsjön the tracer tests did give structural information about zone 2 that was not readily available from the earlier hydraulic testing but the Finnsjön studies also point out a problem that may be inherent in all attempts to use tracer tests for site characterisation: if the test results cannot distinguish between different conceptual process descriptions, can they distinguish between different rock descriptions?

At the WIPP site the tracer tests have been essential for the development of site understanding. The tests were preceded by considerable planning and iteration and they were specially designed to test hypotheses and answer questions. It seems that the strategy has proved fruitful.

In general, tracer tests with conservative tracers can give unique information on flow porosity and tests with slightly sorbing tracers can, additionally, yield information on rock-water-tracer interactions which can hardly be obtained with any other means. However, to fulfil this promise a number of practical difficulties has to be overcome. For example, the information obtained about flow porosity – and flow pathways – is of little value as long as most of the tracers disappear for good. The tests planned to reveal information on interactions can be reliable only if the flow conditions are sufficiently well-known.

A slightly different perspective is to ask what else one could have done: what kind of methods for hydraulic measurement and geophysical studies are available and how much one can deduce from the measured data by different models of interpretation. In the case of alternatives for pumping tracer tests, what is the promise of natural tracer studies? How much can we learn about transport conditions by looking at the geochemistry and mineralogy of the sites?

The apparent difference between natural tracer studies and field tracer tests is that the process that led to the observed distribution of the natural tracer was caused by nature alone and did not result from massive pumping. This is a possible benefit for studies of natural tracer distributions as they are likely to reflect tracer transport in conditions and time scales much more relevant to the safety of waste repositories than the conditions in the pumping tests. Of course, one can argue for the artificial tracer tests that it is the *possible* future flow pathways that we are interested in, but in building a consistent picture of the site the past conditions are the only thing hat we can learn about.

A related potential benefit of natural tracer studies is that these may give us information about the transport in the kind of rock that we would rather wish to have around our repository and not about the high-transmissivity fractures or fracture zones that one usually has to focus on in the pumping tracer tests. Similarly, the scale over which information is obtained is often much larger than in the pumping tracer tests. In Finland work is being done on both the hydrogeological modelling and geochemical evolution modelling at the investigation sites since the late 1980's and gradually some intercomparisons between the interpretations are becoming possible. Often the outcome seems to be on the falsification side: on the basis of the geochemistry certain hydraulic connections can be excluded. For example, in Romuvaara, the chemistry clearly showed that the rock above a certain fracture zone (R9) forms a flow domain separate from the zone and the underlying rock [4]. At

Olkiluoto the distribution and origin of saline and glacial waters offers many challenges to the hydrogeologists and flow modelers. The measurements indicate high salinities starting from the depth of about 500 metres, but, on the other hand, samples from the same depths show traces of glacial melting waters! It is evident that before we are able to use natural tracers effectively in our evaluation we have to understand the processes that led to the present conditions.

4. YES: DO IT BUT THINK IT OVER!

Tracer tests can certainly give useful and sometimes unique information for conceptual modelling and interpretation of some individual actual site features. The tests planned by Posiva in Romuvaara are intended both to help in confirming conceptual assumptions and to scope the utility of tracer tests for possible future characterisation efforts. The idea is to make it an iterative learning exercise where modelling, experimentation and equipment development are tied together in a stepwise working procedure and where before proceeding from any step, an assessment of the progress and potential benefits from continuation can be assessed.

Nevertheless, one may ask us whether or how much the planned tests between two boreholes in Romuvaara contribute to the characterisation of the whole 10 km^2 investigation site. The answer is probably yes and no. Neither the planned tests between two boreholes or any other similar tests could unravel all secrets of the Romuvaara bedrock. It is unrealistic to expect that, by tracer tests or, in fact by any characterisation methods in hand, we could get a complete picture of the transport properties of a given site. The recent developments on Kivetty, another site candidate, give an example on this.

Based on several years of investigations and information from 10 deep boreholes a fairly consistent picture of the major structural features on Kivetty had already evolved by 1995. Then one of the existing boreholes, KR5, was deepened from 500 metres down to 1000 metres and suddenly something like about 100 metres of altered, densely fractured, well-conductive rock was hit. It was evident that, at least at that borehole, it was a major rock feature and, at least locally, it would be important for flow conditions as well. Later a new deep borehole was drilled in the vicinity of KR5 to learn about the spatial extension of the feature, but, to a surprise, no trace of the feature was found in the new hole. Besides, there is another hole near KR5, where no such anomalous rock conditions had been discovered earlier.

Independent of how the new feature finally looks like, it is clear that it can be important for the local transport conditions and it is clear that by any finite surface investigation effort the odds are high that such feature remains undetected. In that sense the tests like those planned for Romuvaara can never yield the whole picture of the volume of heterogeneous rock that is usually studied in site characterisation projects.

On the other hand, we may try to learn as much as possible about the rock that will become the near-field for performance analysis. Direct measurement by tracer tests of the transport parameters for the whole near-field may still be impossible, but these methods can help in abstraction. Once we go underground we should have effective means to decide where we go with our tunnels and where we drill our deposition holes. Flow and transport conditions are one criterion for the discrimination between good and bad rock. Tracer tests can give us valuable information for the basis of such judgements. In this way the tests in Romuvaara may truly help us in the characterisation of the site. We may have to live with the possibility of leaking far-field, but we should try to ensure as good as possible a near-field.

In the past too great expectations may have been attached to tracer tests and their power of proof. The idea of tracer transport simulation may tempt us to believe that we might prove the site performance by direct measurement. That will never be possible for real-life repositories. Tracer tests are one possible means in site characterisation but their application should be judged by their costs and benefits in relation to alternative methods and approaches.

REFERENCES

[1]   Andersson, P., Compilation of tracer tests in fractured rock, 1995, Äspö Hard Rock Laboratory Progress Report 25-95-05, Swedish Nuclear Fuel and Waste Management Co.

[2]   Beauheim, R.L. et al., Rationale for the H-19 and H-11 tracer tests at the WIPP site, Joint NEA/EC Workshop on Field Tracer Transport Experiments, Cologne 28–30 August, 1996.

[3]   Meigs, L. C. et al., Design, modelling and current interpretations of the H-19 and H-11 tracer tests at the WIPP site, Joint NEA/EC Workshop on Field Tracer Transport   Experiments, Cologne 28–30 August, 1996.

[4]   Pitkänen, P. et al., Geochemical modelling study on the age and evolution of the groundwater at the Romuvaara site, 1996, to be published in Posiva Reports Series.

SESSION III

Test Cases: Design, Modelling and Interpretation
Chairmen: J. Vira (Posiva, Finland) and P. Lalieux (NEA)

Field Tracer Transport Experiments at the Site of Canada's Underground Research Laboratory

L.H. Frost, C.C. Davison, T.T. Vandergraaf, N.W. Scheier, and E.T. Kozak
AECL, Whiteshell Laboratories, Pinawa, Manitoba, Canada

Abstract

To gain a better understanding of the processes affecting solute transport in fractured crystalline rock, groundwater tracer experiments are being performed within natural fracture domains and excavation damage zones at various scales at the site of AECL's Underground Research Laboratory (URL). The main objective of these experiments is to develop and demonstrate methods for characterizing the solute transport properties within fractured crystalline rock. Estimates of these properties are in turn being used in AECL's conceptual and numerical models of groundwater flow and solute transport through the geosphere surrounding a nuclear fuel waste disposal vault in plutonic rock of the Canadian Shield.

The different fracture domains at the URL include: fracture zones (faults), defined as volumes of intensely fractured rock; moderately fractured rock, defined as volumes of rock containing a small number of sets of relatively widely spaced, interconnected discrete fractures (joints); and sparsely fractured rock, defined as volumes of rock containing microcracks and very sparsely distributed discrete fractures that are not very interconnected. In addition to natural fracturing, the construction of an underground disposal facility in crystalline rock creates a region of altered stress in the near-field, immediately adjacent to excavated openings. Micro-cracks and small fractures develop in this region and could form additional new pathways for groundwater flow or contaminant transport. The portion of the rock damaged by stress changes due to excavation, or by the excavation method, is referred to as the excavation damage zone. Studies conducted during the construction of the URL facility have shown the extent of these excavation damage zones.

A number of groundwater tracer experiments are currently being performed in the natural fracture domains and excavation damage zones at the URL. These experiments have been used and are being conducted to refine the transport models used in the postclosure environmental and safety assessment of AECL's nuclear fuel waste disposal system and to improve methodologies for future site characterization measurements and experiments. The current status of AECL's groundwater tracer testing program at the URL is described below.

- A series of two-well tracer tests has been conducted within several major low-dipping fracture zones at scales ranging from 17 to 440 m to determine the physical solute transport properties of volumes of intensely fractured rock and to develop methods for extrapolating the test results to larger scales. Based on the successful completion of several smaller-scale tracer tests, this experiment has evolved to larger-scale tests to help establish whether the solute transport properties within zones of intensely fractured rock are scale- or direction-dependent. Currently, a tracer test at a scale of 700 m is in preparation. Equivalent-porous-media models have been used to describe fluid flow and solute advection and dispersion within the fracture zones. Estimates of transport porosity and dispersivity from these models have been used to assign parameter values to the fracture zones in geosphere models for postclosure assessments of AECL's disposal concept.

- In a region of moderately fractured rock containing interconnected networks of discrete fractures, a series of tracer tests is being designed to evaluate the physical and chemical solute transport properties of a relatively large volume (1×10^5 m^3) of such rock. As well, because of the suitability of the porous-media-equivalent method for modelling solute transport in volumes of intensely fractured rock, this modelling approach will also be tested for regions of moderately fractured rock during this experiment. Other modelling approaches such as discrete fracture network models, will also be evaluated. The geological, geophysical, geochemical and hydraulic characterization of this region is currently underway.

- A tracer experiment has recently been conducted within a region of the excavation damage zone in the floor of a 3.5 m diameter test tunnel located on the 420 Level of the URL facility to obtain information on the physical solute transport properties within excavation damage zones surrounding underground tunnels. Both mass flux and analytical modelling calculations were performed to determine the permeability, transport porosity and dispersivity characteristics of the excavation damage zone. This initial experiment was performed to support the development of vault and geosphere models for a postclosure assessment of the in-room emplacement of copper containers of used CANDU fuel in a hypothetical permeable geosphere. This assessment complemented an earlier postclosure assessment of borehole emplacement of titanium containers in a region of very low permeability rock. Further tests of this type are planned to be incorporated into the excavation stage of the moderately fractured rock experiment.

- A migration experiment has been designed, in cooperation with the Japan Atomic Energy Research Institute (JAERI), to study the transport of conservative and sorbing radionuclides in natural fractures in excavated blocks of granite (approximately 1 m^3 in size) under in-situ geochemical groundwater conditions. This experiment is being conducted in an IAEA Class B laboratory specifically constructed for this purpose at the 240 Level of the URL using rock blocks containing natural fractures excavated from a nearby vertical fracture.

INTRODUCTION

AECL is conducting hydrogeological research at its Underground Research Laboratory (URL) in southeastern Manitoba, Canada, as part of its evaluation of the concept for disposal of nuclear fuel waste deep in plutonic rock masses of the Canadian Shield. The feasibility of this concept and assessments of its impact on the environment and human health are documented in an Environmental Impact Statement that was submitted for public, regulatory and scientific review [1],[2].

The primary objective of AECL's hydrogeological research is to develop and demonstrate methods for determining the chemical and physical characteristics of groundwater flow systems in plutonic rock bodies of the Canadian Shield at the various size scales relevant to the prediction of potential radionuclide migration through the geosphere surrounding a disposal vault constructed at a depth of 500 to 1 000 m. Investigations conducted as part of this research indicate that the degree of fracturing in crystalline rock of the Canadian Shield is one of the primary distinguishing features between volumes of rock that have significantly different groundwater flow and solute transport characteristics [3]. The different fracture domains identified in crystalline rock of the Shield are:

- fracture zones (faults), which are volumes of intensely fractured rock;
- moderately fractured rock, which are volumes of rock containing a small number of sets of relatively widely spaced, interconnected discrete fractures (joints); and,
- sparsely fractured rock, which are volumes of rock containing microcracks and very sparsely distributed discrete fractures that are not very interconnected.

All three of these fracture domains exist within the upper 500 m of rock mass at the site of AECL's URL (Figure 1).

In addition to natural fracturing, the construction of an underground disposal vault will create a region of altered stress in the near-field, immediately adjacent to the excavated openings where micro-cracks and small fractures will develop. The portion of the rock damaged by stress changes due to excavation, or by the excavation method, is referred to as the excavation damage zone (EDZ). Studies conducted during the construction of the URL have shown the extent of these EDZ's [4]. It is expected that the EDZ will have properties that are considerably different from those of the undisturbed rock mass, such that the EDZ might provide additional new pathways for groundwater flow or contaminant transport from the vault.

This report provides a brief summary of the groundwater tracer experiments that have been, or are being, performed in the natural fracture domains and excavation damage zones at the URL. These experiments have been conducted to refine the transport models used in the postclosure environmental and safety assessment of AECL's nuclear fuel waste disposal system and to improve methodologies for future measurements and experiments.

THE UNDERGROUND RESEARCH LABORATORY

The URL constitutes a well-characterized in situ environment in a previously undisturbed volume of rock for experiments to address various geotechnical and engineering issues of importance to AECL's disposal concept and to demonstrate design and engineering elements of the proposed disposal concept. Prior to any excavation at the URL site, a detailed site evaluation program was carried out to characterize the geological, hydrogeological and geochemical conditions of the 3.8 km^2 area of the site to a depth of approximately 500 m. The information from this program was used to select the location for the URL shaft and the underground facility so as to provide access to different fractured rock domains and lithologies in the granitic Lac du Bonnet batholith. The underground workings of the URL comprise a vertical access shaft to a depth of 443 m and major testing levels at depths of 240 and 420 m (each comprising several hundred meters of tunnels) as well as a ventilation shaft connecting the testing levels to the surface (Figure 2).

TEST CASES: DESIGN, MODELLING AND INTERPRETATION

FRACTURE ZONES

An initial series of seven two-well tracer tests (phase 1) has been conducted within a major low-dipping fracture zone (Fracture Zone 2–Figure 1) in the rock mass at the URL site at scales ranging from 17 to 209 m to determine the physical solute transport properties of volumes of intensely fractured rock [5]. The two-well tracer tests have been performed as steady state, recirculating tests: conservative groundwater tracers were injected as a pulse source into a continuous withdrawal/injection flow field, and tracer concentrations of the recirculated water were monitored throughout the duration of each test (Figure 3). To help evaluate the results of this first phase of testing, a series of six additional, smaller-scale tests (phase 2) were planned and conducted within a high permeability region of Fracture Zone 2 to investigate different groundwater tracer testing techniques [6]. These tests were designed to: compare two-well recirculating, non-recirculating and convergent tracer test techniques; examine the effect of direction on the tests; examine the effects of different flow rates or pressure gradients; and to compare the transport behavior of different tracers including different anionic tracers, colloid tracers and redox-controlled chemical tracers. Based on the successful completion of phases 1 and 2, this experiment evolved to larger-scale tests (phase 3) to help establish whether the solute transport properties within zones of intensely fractured rock are scale- or direction-dependent and to develop methods for extrapolating the test results to larger scales by combining two-well tracer tests with crosshole hydraulic response tests. As part of the phase 3 testing, a two-well tracer test has been conducted at a scale of 440 m within Fracture Zone 1.5 and a two-well tracer test at a scale of 700 m within Fracture Zone 1 is in preparation (Figure 1).

The suitability of the porous-media-equivalent method in describing fluid flow and solute transport within fracture zones has also been evaluated during this study. Because of the apparent inhomogeneity of the flow and transport properties of the fracture zones, and the recirculating mode of the two-well tests, the finite-element computer code MOTIF (Model of Transport in Fractured/ Porous Media) developed by AECL [7], has been used to solve the fluid flow and solute transport equations. MOTIF allows the fracture zones to be represented using quadrilateral elements of differing thickness, permeability, porosity and orientation in 3-D space. This modelling approach has been used successfully to simulate flow and radiotracer transport in a similar fracture zone at the nearby Whiteshell Laboratories site [8].

Well-defined tracer breakthroughs curves were obtained for all of the two-well tracer tests that have been performed so far at scales ranging from 17 to 440 m. At a transport scale of less than about 20 m, double, sharp, tracer peaks were observed in the breakthrough curves. The second peaks may be due to tracer recirculation or flow channelling or some combination of both. At a transport scale greater than about 30 m, single, well-dispersed tracer concentration peaks were observed. These results suggest that, at the larger transport scales, there is an averaging of the solute transport properties within these fracture zones.

Modelling of the individual tracer tests indicates that the porous-media-equivalent method is suitable for modelling fluid flow and solute transport in these types of fracture zones. Adequate fluid flow models have been developed for the regions of the fracture zones tested so far. However, both small- and large-scale permeability variations within the fracture zones must be accounted for in the analysis. Development of a suitable transport model capable of simultaneously describing all the tracer tests conducted within the same fracture zone has been more difficult. Simulations of tracer transport using models with uniform transport properties do not describe the test results very well. However, initial simulations with a model having varying "effective thickness" (product of transport porosity and fracture zone thickness) indicate that this type of model may reasonably describe all of the test results. Further work is required in order to develop a model that can simulate all the tracer tests conducted during these different phases of tracer testing.

EXCAVATION DAMAGE ZONE

A tracer experiment has recently been conducted within a region of the excavation damage zone (EDZ) in the floor of a 3.5 m diameter test tunnel to obtain information on the physical solute transport properties within EDZ's surrounding underground tunnels [9]. The tunnel was constructed as part of AECL's Mine-by Experiment to investigate the formation and geomechanical characteristics of the EDZ adjacent to an underground opening. A key goal in the design of the Mine-by Experiment was to conduct the investigation in a geological/geotechnical environment similar to that which might be expected between a depth of 500 and 1000 m in the Canadian Shield. To achieve this, the tunnel was located on the 420 Level of the URL (Figure 2) where the stress conditions are similar to those at a depth of about 1000 m in other parts of the Shield [10].

Following completion of the test tunnel, which was constructed using a mechanical excavation technique to avoid the damage effects of blasting, the connected permeability test phase of the Mine-by Experiment was performed to characterize the groundwater flow properties of the EDZ located in the floor of the tunnel [4]. The results from this phase of testing indicated that the main flow pathway within the EDZ in the floor of the tunnel is within a process zone of intense fracturing (Figure 4); virtually no flow occurred outside the region of the process zone. From the point of view of solute transport, the region of the process zone is a potential pathway for contaminant migration within the tunnel.

The EDZ tracer experiment was performed as a constant head test by continuously injecting a constant concentration of conservative tracer into a region of the process zone, and monitoring tracer breakthrough from the zone at a distance 1.5 m away. The results obtained during the test show good tracer breakthrough. An equivalent-porous-media approach was used in analyzing fluid flow and solute transport within the process zone. Both mass flux and analytical modelling calculations were performed to determine the permeability, transport porosity and dispersivity characteristics of the zone. The results from this initial experiment have been used to support the development of vault and geosphere models for a postclosure assessment of the in-room emplacement of copper containers of used CANDU fuel in a hypothetical permeable geosphere. This assessment complemented an earlier postclosure assessment of borehole emplacement of titanium containers in a vault situated in a geosphere region of very low permeability rock. Further tests of this type are planned to be incorporated into the excavation stage of the moderately fractured rock experiment which is described below.

PLANNED TRACER TRANSPORT EXPERIMENTS

Tracer transport experiments are currently planned for three distinctly different fractured rock domains at the URL. These are described below.

FRACTURE ZONES

Final preparations have been completed for a two-well tracer test at a scale of 700 m within Fracture Zone 1 (Figure 1). This test is being performed as part of the phase 3 series of two-well tracer tests within major low-dipping fracture zones to determine the physical solute transport properties of volumes of intensely fractured rock and to develop methods for extrapolating the test results to larger scales. The results of previous tracer tests, conducted as two-well, steady state, recirculating tests at scales ranging from 17 to 440 m, along with data from hydraulic tests conducted in the same pairs of boreholes, indicate that a relationship may exist between values of transport porosity determined from tracer tests and storativity determined from hydraulic tests. It appears that the results of large-scale hydraulic tests combined with smaller-scale hydraulic tests and two-well tracer tests could be used to estimate the transport porosity of large areas of major fracture zones in a plutonic rock mass. This test is being performed to address this relationship and if successful, it may eliminate the need to perform many expensive, large-scale tracer tests in major fracture zones during the site evaluation phases of a disposal vault siting project.

MODERATELY FRACTURED ROCK

A series of tracer tests are being planned for a region of moderately fractured rock (MFR) on the 240 Level of the URL (Figure 2) to evaluate the physical and chemical solute transport properties of a large volume of MFR and to determine the suitability of the porous-media-equivalent method for modelling solute transport in regions of MFR. For the purposes of this experiment moderately fractured rock is defined as a region of rock mass at least 100,000 m^3 in size having a relatively uniform distribution of intersecting permeable fractures and a fracture frequency (based on a line sample) of 1 to 5 fractures per meter.

The region of MFR has been defined by a series of boreholes drilled from the underground workings of the URL and from surface (Figure 5). This region is transected by a splay of a major low dipping fracture zone (Fracture Zone 2.5–Figure 1). A north-east/north-northwest trending set of near vertical fractures also extends through this region of the rock mass. The volume of interest for the MFR transport experiment measures roughly 50 m x 50 m x 50 m.

Three phases of work are currently planned for the MFR tracer experiment. Phase 1 which is nearing completion has involved excavation of the access tunnel to the region of MFR and the drilling of additional boreholes for the geological, geophysical, geochemical and hydraulic characterization of the test region. The characterization of the test region is near completion. An initial tracer test will be performed during phase 1 at a scale in the range of 10 to 50 m. Many aspects of the design for this test have been based on the successful completion of the series of two-well tracer tests performed within the major low-dipping fracture zones. It is planned to perform the tracer test as a two-well, steady state, non-recirculating test. The testing method, procedures and equipment that will be used during subsequent tracer tests for this experiment will be evaluated following the completion of this initial tracer test. Phase 2 will involve activities related to conducting a series of groundwater tracer tests using both non-reactive and reactive tracers. Phase 3 will involve sampling the fracture network to determine tracer channelling and the distribution of reactive tracers using either a network of boreholes or by excavating a test tunnel through the central portion of the experimental area after completion of phase 2. If a tunnel is excavated through the moderately fractured rock area a series of radial-convergent conservative and non-conservative groundwater tracer tests may be planned for this last phase of work as well as tracer tests to determine the solute transport properties of the EDZ.

NATURAL FRACTURE IN SPARSELY FRACTURED ROCK

An underground radioisotope laboratory has been designed and constructed at the 240 Level of the URL (Figure 2) to study the transport of conservative and sorbing radionuclides in natural fractures under in situ geochemical conditions [11]. This experiment has been designed in cooperation with the Japan Atomic Energy Research Institute (JAERI).

The migration experiment is being performed in an IAEA Class B radioisotope laboratory which is located adjacent to a subvertical joint zone, the only permeable fracture intersecting the URL at this depth. Blocks of granite, with dimensions of about 1 m x 1 m x .7 m, each containing a natural fracture, have been excavated from this joint zone using a diamond wire saw cutting technique (Figure 6). Groundwater from the joint zone was used as drilling and cutting fluid to minimize contamination of the fractures.

Two blocks of granite are being used for the migration experiments; each block has been equipped with 12 inlet/outlet ports at the periphery of the fractures. These ports have been used to hydraulically characterize the fractures to select the most appropriate flow path for the migration experiments. The migration experiments will be performed using groundwater from the joint zone as the transport solution to maintain the low Eh, in situ, geochemical conditions within the natural fractures. Initial migration experiments have been completed using bromide as a conservative tracer at flow rates ranging from 5 to 400 mL/h. These experiments will be followed by migration experiments using sorbing radioisotope tracers and colloidal material. At the completion of the migration experiments, the blocks will be separated at the fracture and the surfaces analyzed radiometrically to provide further information on the transport pathways within the natural fractures.

ACKNOWLEDGMENTS

The Canadian Used Fuel Disposal Program is funded jointly by AECL and Ontario Hydro under the auspices of the CANDU Owners Group. The program on radionuclide migration experiments under in situ conditions is being funded by JAERI under a cooperative agreement.

REFERENCES

[1]   AECL, Environmental impact statement on the concept for disposal of Canada's nuclear fuel waste, Atomic Energy of Canada Limited Technical Report, AECL-10711, COG-93-1, 1994.

[2]   AECL, Summary of the environmental impact statement on the concept for disposal of Canada's nuclear fuel waste, Atomic Energy of Canada Limited Technical Report, AECL-10721, COG-93-11, 1994.

[3]   Davison, C.C., Brown, A., Everitt, R.A., Gascoyne, M., Kozak, E.T., Lodha, G.S., Martin, C.D., Soonawala, N.M., Stevenson, D.R., Thorne, G.A. and Whitaker, S.H., The disposal of Canada's nuclear fuel waste: Site screening and site evaluation technology, Atomic Energy of Canada Limited Report, AECL-10713, COG-93-3, 1994.

[4]   Chandler, N.A., Kozak, E.T. and Martin, C.D., Connected pathways in the EDZ and the potential for flow along tunnels, In: Proceedings of the International Conference on Deep Geological Disposal of Radioactive Waste - EDZ Workshop, Winnipeg, Manitoba, Canada, 1996 September 20, 25-34.

[5]   Frost, L.H., Scheier, N.W. and Davison, C.C., Transport in highly fractured rock experiment - Phase 1 tracer tests in Fracture Zone 2, Atomic Energy of Canada Limited Technical Record, TR-672, COG-95-85, 1995.

[6]   Frost, L.H., Scheier, N.W. and Davison, C.C., Transport in highly fractured rock experiment - Phase 2 tracer tests in Fracture Zone 2, Atomic Energy of Canada Limited Technical Record, TR-685, COG-95-198, 1995.

[7]   Chan, T., Scheier, N.W. and Reid, J.A.K., Finite-element thermohydro-geological modelling for Canadian nuclear fuel waste management, Proceedings of the Canadian Nuclear Society, Second International Conference on Radioactive Waste Management, Winnipeg, Canada, 1986, 653-660.

[8]   Frost, L.H., Scheier, N.W. and Davison, C.C., Two-well radioactive tracer experiment in a major fracture zone in granite, Atomic Energy of Canada Limited Technical Record, TR-671, 1995.

[9]   Frost, L.H. and Everitt, R.A., Excavation damage zone tracer experiment in the floor of the Room 415 test tunnel, Atomic Energy of Canada Limited Report, AECL-11640, COG-96-321, 1996.

[10]   Read, R.S. and Martin, C.D., Technical summary of AECL's mine-by experiment phase: excavation response, Atomic Energy of Canada Limited Report, AECL-11311, COG-95-171, 1996.

[11]   Vandergraaf , T.T., Drew, D.J., Kumata, M. and Nakayama, S., Design, construction and operation of an underground facility to study the migration of radioisotopes in natural fractures under in situ conditions, In: Extended Abstracts of the 4th International Conference on Nuclear and Radiochemistry (F. David and J.C. Krupka, editors), St. Malo, France, 1996 September 8-13, paper E-P40.

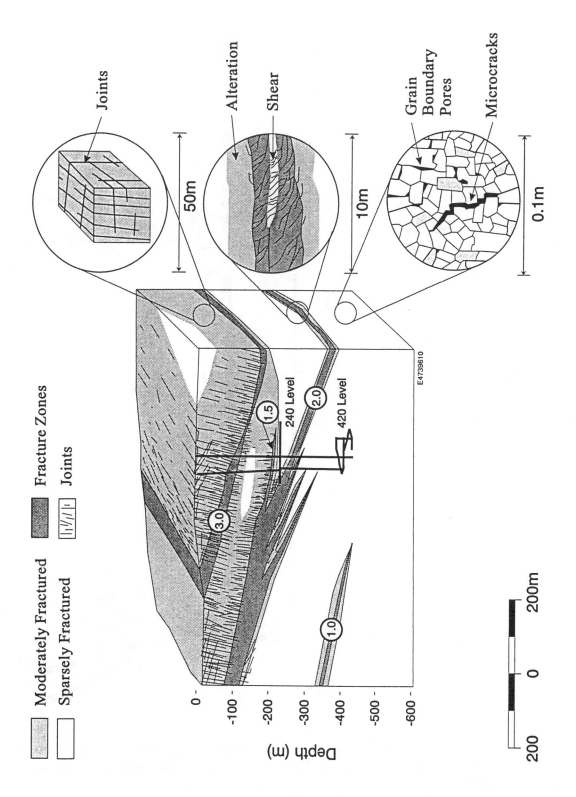

Figure 1. Geologic cross section showing the different natural fracture domains at the URL: fracture zones, moderately fractured rock and sparsely fractured rock

139

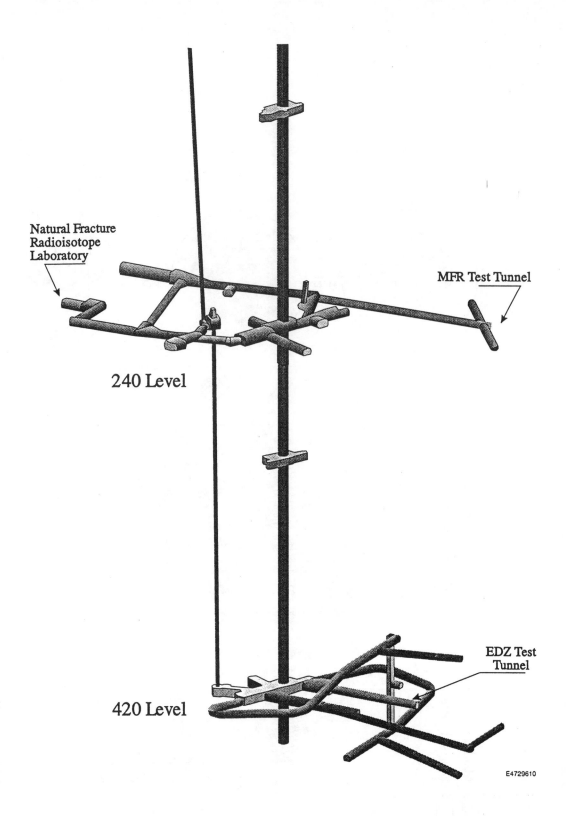

Natural Fracture
Radioisotope
Laboratory

MFR Test Tunnel

240 Level

EDZ Test
Tunnel

420 Level

E4729610

Figure 2. Sectional view of the URL. Location of the excavation damage zone (EDZ), moderately fractured rock (MFR) and natural fracture transport experiments are shown.

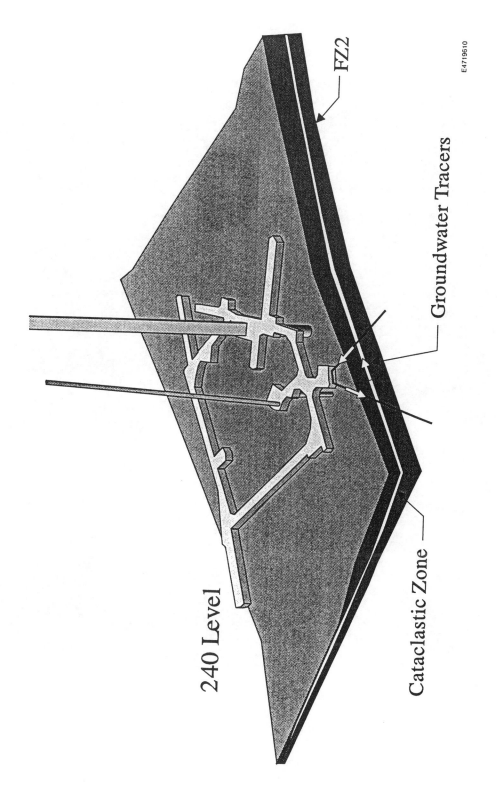

Figure 3. Schematic of the two-well tracer testing technique at the URL

Fig. 4: Photo of the process zone of intense fracturing located in the EDZ of the test tunnel floor

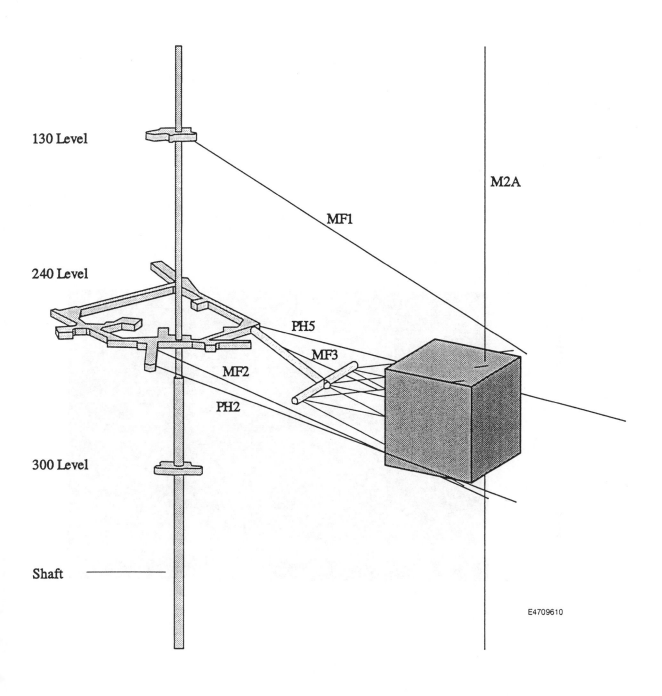

130 Level

M2A

MF1

240 Level

PH5

MF3

MF2

PH2

300 Level

Shaft

E4709610

Figure 5. MFR experimental layout

Figure 6. Block of excavated granite containing a natural fracture

The Äspö TRUE experiments

Olle Olsson[1] and Anders Winberg[2]

[1] SKB Äspö Hard Rock Laboratory, Sweden, [2] Conterra AB, Göteborg, Sweden

Abstract

SKB's concept on deep geological disposal of spent nuclear fuel is based on a multi-barrier system for isolation of the spent fuel from the biosphere. The barriers are a low-solubility waste form, encapsulation of the fuel in a copper canister, a bentonite buffer surrounding the canister, and the host rock. In case of an initial canister damage, the retention capacity of the host rock for the short lived radionuclides such as Cs and Sr is important. For the Operating Phase of the Äspö Hard Rock Laboratory the need for a better understanding of radionuclide transport and retention processes was recognized. This included enhancement of confidence in models to be used for quantifying transport of sorbing radionuclides in performance assessment. Further, to be able to show that pertinent transport data could be obtained from site characterization or field experiments and that laboratory results could be related to retention parameters obtained *in situ*. To resolve these issues SKB initiated the Tracer Retention Understanding Experiments (TRUE). The TRUE program is based on a staged approach where data for conceptual and numerical modelling should be provided at regular intervals. Periodic evaluation of test results and a close integration of experimental and modelling work should provide the basis for detailed planning of subsequent test cycles and successive improvement of models. The planned duration of each stage is approximately 2½ years with a total duration of the program of about 10 years, beginning in 1994 and ending in 2003. The basic idea is to perform a series of tracer experiments with successively increasing complexity. Each tracer experiment will consist of a cycle of activities beginning with geological characterization, modelling, followed by a set of hydraulic and tracer tests. Finally resin will be injected and the tested rock volume excavated, and analyzed for flow path geometry and tracer concentration. The first test cycle, TRUE-1, which currently is in progress, is small scale, of limited time duration, and primarily aimed at technology development though tests with sorbing tracers will conclude the test cycle. Supporting work include development of techniques for pore space characterization using resin and batch sorption and through diffusion tests using weakly sorbing tracer tests in the laboratory.

Background

SKB's concept on deep geological disposal of spent nuclear fuel is based on a multi-barrier system for isolation of the spent fuel from the biosphere. The barriers are a low-solubility waste form, encapsulation of the fuel in a copper canister, a bentonite buffer surrounding the canister, and the host rock. The host rock should provide a stable mechanical and chemical environment for the engineered barriers and it should reduce and retard transport of radionuclides released from the engineered barriers. In case of an initial canister damage, the retention capacity of the host rock for the short lived radionuclides such as Cs and Sr is important. Actinides become important in a longer time perspective.

In planning the experiments to be performed during the Operating Phase of the Äspö Hard Rock Laboratory the need for a better understanding of radionuclide transport and retention processes was recognized. The needs of performance assessment included increased confidence in models to be used for quantifying transport of sorbing radionuclides. It was also considered important from the PA perspective to be able to show that pertinent transport data (such as flow wetted surface, K_d, diffusivity, etc.) could be obtained from site characterization or field experiments and that laboratory results could be related to retention parameters obtained *in situ*.

The initial planning resulted in three proposed experiments (Pore Volume Characterization, Multiple Well Tracer Experiment, and Matrix Diffusion Experiment) to study transport processes in the detailed scale (<10 m) and one experiment for studies in the block scale (10-100 m). The proposals were then submitted for review. The reviewers suggested modifications of, and additions to the MWTE and MDE experiments which made both experiments very comprehensive undertakings. The additions also made the experiments quite similar. SKB then decided to merge the initially planned experiments into a common tracer test program named the Tracer Retention Understanding Experiments (TRUE).

Scope and objectives

The objectives of TRUE are:

- Develop the understanding of radionuclide migration and retention in fractured rock.
- Evaluate to what extent concepts used in models are based on realistic descriptions of rock and if adequate data can be collected in site characterization.
- Evaluate the usefulness and feasibility of different approaches to model radionuclide migration and retention.
- Provide *in-situ* data on radionuclide migration and retention.

The TRUE program is based on a staged approach where data for conceptual and numerical modelling should be provided at regular intervals. Periodic evaluation of test results and a close integration of experimental and modelling work should provide the basis for detailed planning of subsequent test cycles and successive improvement of models. Detailed plans for the planned tracer test cycles will be developed successively during the execution of this experimental program. The planned duration of each stage is approximately 2½ years and total duration of the program is nearly 10 years, beginning in 1994 and ending in 2003.

The basic idea behind TRUE is to perform a series of tracer experiments with successively increasing complexity. In principle, each tracer experiment will consist of a cycle of activities beginning with geological characterization of the site, modelling, followed by a set of hydraulic and tracer tests, finally resin will be injected, the tested rock volume excavated, and analyzed for flow path geometry and tracer concentration. The first test cycle, which currently is in progress, is small scale, of limited time duration, and primarily aimed at technology development. The initial tests will be followed by detailed scale tracer tests of longer time duration which allow tests of retention mechanisms. A block scale tracer test will be initiated and detailed scale tracer tests will be performed within the rock volume of the block scale tracer tests. This should provide a basis for understanding of scaling relationships and a test of modelling capabilities for radionuclide transport at the 50 m scale. In addition to the field experiments, the TRUE program includes supporting development of tracer test technology and laboratory investigations of sorption and diffusion data for selected radionuclides.

The Äspö Task Force on Modelling of Groundwater Flow and Transport of Solutes has been engaged for providing advice on experimental design, predictive modelling, and evaluation of experimental data from TRUE. The close interaction between TRUE and the Task Force is foreseen to become an important element in the evaluation of different conceptual models of radionuclide transport.

TRUE Stage 1 - Current results

The first stage of TRUE (TRUE-1) is aimed at understanding tracer transport in a single feature, which could represent a fracture intersecting a canister deposition hole. TRUE-1 was initially aimed at test of equipment, adaption of tracer test methodology to Äspö conditions, and understanding of transport of conservative tracers. Later the program was expanded to include also field tests with weakly sorbing tracers. In addition a technology for obtaining the internal structure of pore space in the fracture is developed. The first stage will be concluded by injection of epoxy resin, followed by excavation (large diameter drilling) and subsequent analysis of the pore space. Within the project the role of microbes in mass transfer will also be addressed.

As indicated above, different tracer test methodologies will be tested and evaluated within TRUE-1. The defined tracer test programme includes the following components;

- Preliminary tracer tests
- Radially converging tracer test
- Dipole tracer tests
- Complementary tracer tests
- Tracer test/-s with weakly sorbing tracers

Before onset of the tests a suitable experimental site had to be identified that met the specifications set up in advance. These specifications included i.a. homogeneous lithology, low fracture frequency and well-defined fracture sets, isolated rock volume, relative hydraulic isolation, reducing groundwater chemistry and a transmissivity T of the target feature ~10^{-7} m^2/s. A site selection programme was initiated to identify suitable experimental sites for TRUE and other experiments to be performed within the Operational Phase of the Äspö HRL. It was identified that sites for the planned experiments had to be sited using hard data from pilot boreholes. It was also identified that the experimental volumes should be located outside the tunnel spiral, partly to avoid complicated boundary conditions, but also because of noted enhanced connectivity within the spiral. By concentrating on the rock volume east of the last spiral turn, c.f. Figure 1, it was possible to utilize the identified highly conductive fracture zone NNW-4 as moderator for hydraulic disturbances. The defined site selection programme, called SELECT, involved drilling of eight 56 mm cored boreholes, 15-70 m long, with the objective of finding suitable locations for the TRUE, REX and Chemlab experiments. The boreholes /cores were subject to;

- core logging
- borehole TV (BIPS)
- pressure build-up tests in selected intervals
- installation of downhole packer assemblies
- performance of an interference test programme

-acoustic flow logging (UCM)
- directional borehole radar
- water sampling
- monitoring of pressure over time

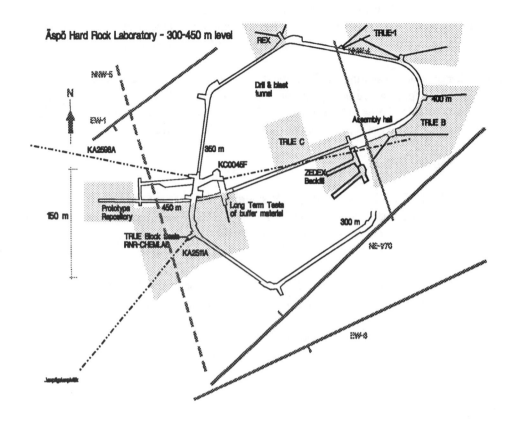

Figure 1   **Äspö Hard Rock Laboratory - Overview of experimental level showing major structural features, the location of TRUE-1 and other ongoing and planned experiments**

The results from the interference test programme was used to test connectivity and potential for disturbance between different potential experimental volumes. All parts of the drilling equipment and all downhole logging tools were cleaned using hot-water high-pressure cleaner to minimize introduction of bacteria into the boreholes. The collected characterization data was used to assess whether suitable target features for the experiments were available, and to what extent the identified experimental volumes and target features were likely to be affected by activities in neighbouring rock blocks.

Three features were identified as suitable for further study within TRUE-1 in borehole KA3005A. To investigate these features further, a new niche was developed from which an additional four cored 56mm boreholes, KXTT1-KXTT4, were drilled to intercept all three features located some 10-15 m into the rock. Pressure responses in the instrumented borehole KA3005A and temporary downhole installations in the new boreholes were used for preliminary assessments of connectivity and preliminary structural inferences. Basically, the investigation methods mentioned above were utilized in the characterization in the boreholes making up the TRUE-1 borehole array. Since degassing phenomena had partially hampered the acoustic flow logging during SELECT, a single packer flow logging method was applied instead which allowed a characterization of inflows with a 0.5 m resolution, c.f. Figure 2. Following the basic characterization, the boreholes were equipped with packer systems with up to five test sections followed by an elaborate interference test programme. Response tables based on defined indices were utilized in interpreting the response patterns and degree of reciprocity in the noted responses. In order to provide conservative transport parameters (transport porosity, dispersivities) for model predictions of the planned radially converging tracer

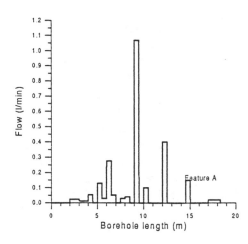

Figure 2 Results of single packer flow logging in borehole KXTT2

tests, two preliminary tracer tests were performed in Features A and B, respectively, using a radially converging flow field established between two borehole sections.

Subsequently, data from the performed characterization were integrated and used to build descriptive models on detailed and block scales using both stochastic continuum and discrete fracture network . The building of descriptive models was a joint effort involving the SKB TRUE Project team and teams from USDOE/LBNL and PNC/Golder [1], [2]. The resulting deterministic structural-hydraulic model of the TRUE-1 Block is shown in Figure 3. PNC/Golder produced a discrete fracture network model based on the collected data [2]. Based on the analysis of the characterization data and the developed descriptive models, one of the identified

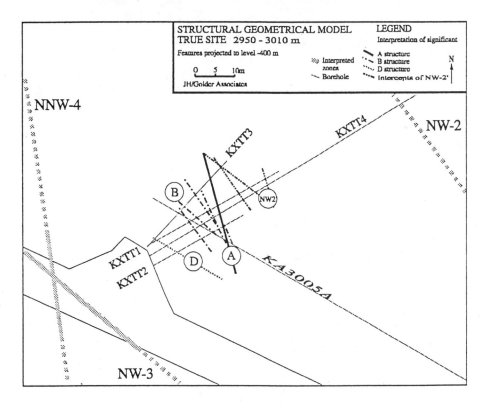

Figure 3 Structural-geological model of the TRUE-1 site. Horizontal section at Z=-400 masl showing bounding minor fracture zones and identified 4th order features.

potential target features, Feature A, was proposed for further study. This feature is a reactivated mylonite, a representative conductor at Äspö, with a transmissivity in the order of $0.1\text{-}3\cdot10^{-7}$ m²/s as observed from the flow and pressure build-up tests. The interpretation of the interference tests indicated a somewhat higher transmissivity, ranging from $7\cdot10^{-9}$ - $5\cdot10^{-7}$ m²/s. The performed hydraulic tests indicate pseudo-spherical responses throughout the array, indicating three-dimensional connectivity. However, monitoring of steady state pressure indicate that Feature A, with one exception, is relatively well isolated from neighbouring hydraulic conductors. Hence, Feature A has been interpreted as being part of a leaky aquifer system and the interference tests have been interpreted accordingly [1]. This latter conceptual standpoint was derived late in the Spring 1996 and was not conveyed to the modelling teams of the Task Force performing predictions of the radially converging tracer test. The results of the preliminary tracer test in Feature A indicate a flow porosity of $1.8\cdot10^{-4}$ and a dispersivity of 0.6 m. The variation in hydraulic head in the array is about 1m over a length scale of 10m, ie. a hydraulic gradient of 10% in Feature A within the array. An interpolated picture of the hydraulic head situation prior to the radially converging tracer test (RC-1) is shown in Figure 4.

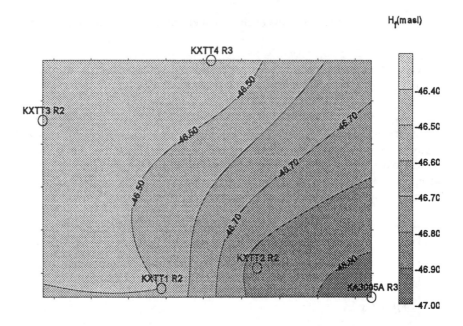

Figure 4 Hydraulic head distribution in the plane of Feature A (SURFER interpolation) prior to the start of the TRUE-1 RC-1 radially converging tracer test (January 17th, 1996).

The modelling teams in the Äspö Task Force implemented the reported data and descriptive models to make their blind predictions of the planned radially converging tracer test (RC-1). The purpose of the RC-1 experiment was to determine transport parameters of Feature A and to test equipment and conservative tracers to be employed in subsequent tests and stages of TRUE. Various continuum model approaches (deterministic and stochastic, analytical and numeric), discrete fracture network and channel network model approaches were used by the modelling teams. The RC-1 experiment involved establishment of a constant flow of 200 ml/min in a section in borehole KXTT3 containing Feature A and subsequent injection of decaying tracer pulses in four sections in the remaining boreholes intercepting Feature A, located 4-10 m away from the pumped section, c.f. Figure 5. Details on the respective injections are shown in Table 1. The experiment showed breakthrough from the two injection sections in KXTT1 and KXTT4 with mass recoveries of 91 and 97%, respectively. The breakthroughs showed little or no effects of processes, i.e. the breakthroughs are simply translations in time of the injection signal, c.f. Figure 6. No breakthrough was observed from the remaining two injection sections in KXTT2 and KA3005A, which are located farthest away from the pump hole, and which also are located closest to the tunnel. A subsequent stepwise increase in the pump flow to

Borehole	Section	Tracers	Inj. conc C_{00} (mg/l)	Max conc C_0 (mg/l)
KXTT1	R2	Uranine	5000	42
		Gd-DTPA	5000	42
KXTT2	R2	Rhodamine WT	8750	73
		Eu-DTPA	3000	25
KXTT4	R3	Amino G Acid	9000	450
		Ho-DTPA	1000	50
KA3005A	R3	Eosin Y	13000	130
		Tb-DTPA	9000	90

Table 1. Details on tracer injections made during TRUE-1 RC-1.

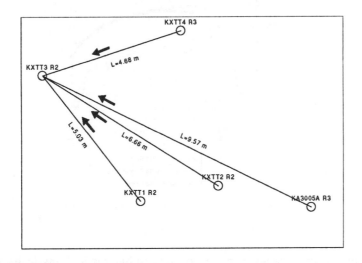

Figure 5 TRUE-1 RC-1 - Testgeometry and pattern of borehole intersections in the plane of Feature A.

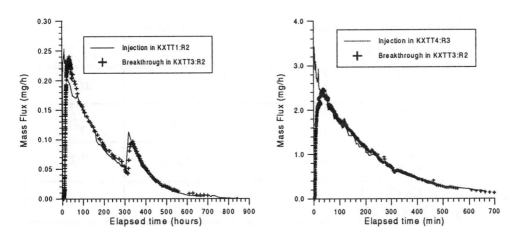

Figure 6   Comparison between concentration of tracer in the injection section in
a) KXTT1:R2 (Uranine), b) KXTT4:R4 (Amino G) with the corresponding
breakthrough concentration in the pumped section KXTT3:R2.

400ml/min and finally 3.4 l/min, enabled breakthrough from the remaining two injection sections, but with very low mass recoveries. A preliminary comparison between model predictions and experimental results show that most modelling teams predicted breakthrough from all four injections, although some teams predicted distinctly lower mass recoveries from the two injections which *in-situ* did not produce breakthrough. The SKB TRUE Project team predicted breakthrough for all four injections at the initial pump rate. Figure 7 shows that the agreement between predicted and experimental breakthrough is quite good. The breakthrough times predicted by the modelling teams are also in accord with those observed in the experimental results. The latter can to a large extent be explained by the calibration of transmissivity and flow porosity against the performed preliminary tracer tests, one of which utilized a flow path successfully tested in RC-1. From the predictions, performance measures have been calculated for breakthrough time t_5, t_{50}, t_{95}, defined as the time when 5, 50, and 95% of the mass has been recovered. Table 2 compares the prediction results of the TRUE team with the experimental results. Most modelling teams overpredicted the

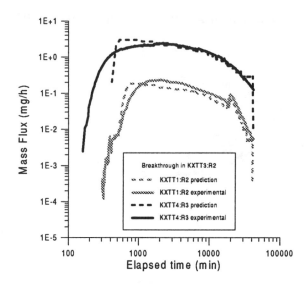

Figure 7 Comparison between predictions (ensemble results) made by the SKB TRUE Project team and the corresponding experimental results (log-log representation)

Injection section	t_5 (hours)	t_{50} (hours)	t_{95} (hours)
KXTT1:R2 Prediction[1]	19.6	150.2	463.5
KXTT1 R2 Experimental	25.5	154	488
KXTT4:R3 Prediction[1]	16.5	151.9	592.5
KXTT4:R3 Experimental	20.9	156	560

Table 2 Tracer travel times for TRUE-1 RC-1. [1] Predictions conditioned on measured transmissivities and *in-situ* steady state head data

drawdown in the injection sections resulting from the pumping, which show that the confined aquifer representations commonly used do not reflect the leaky aquifer behaviour attributed to Feature A. Subsequent to the RC-1, a series of four dipole experiments (DP1-DP4) have been performed. The objective of these tests being to tests dipole methodology and equipment in low-transmissive features and to compare the obtained results with the RC-1 results. These dipole tests are also subject to predictions by the Äspö Task Force. Recently, three complementary tests were performed to assist in locating the planned test with sorbing tracers. Evaluation of the performed tracer tests is in progress. In parallel, a parameter estimation exercise integrating all transient interference tests data is under way with the aim to test the present descriptive model, and suggest alternative structural models.

Supporting work

Batch sorption and through diffusion experiments are performed on generic Äspö material and site specific rock material in order to a) test suitable slightly sorbing (reactive) tracers, b) obtain laboratory derived transport parameters for non-reactive and reactive tracers (K_d and diffusivities), and c) test sorption of reactive tracers on equipment parts used in the field tracer tests. Results for the Äspö materials investigated so far indicate that K_d for the tracers investigated range from 10^{-6} -10^{-1} m³/kg, with the following relative magnitude;

$$^{22}Na^+ < {}^{47}Ca^{2+} \approx {}^{85}Sr^{2+} << {}^{86}Rb^+ \approx {}^{133}Ba^{2+} < {}^{137}Cs^+$$

The diffusivity experiments are performed on material with three different thicknesses (10, 20 and 40 mm). Results on generic Äspö diorite show effective diffusivities D_e in the order of $5\text{-}8\cdot10^{-14}$ m²/s. Further evaluation of tests on the above tracers are in progress.

Preliminary results from the test of sorption on *in-situ* equipment indicate that brass fittings may act as a sink for sorbing tracers .

As a trial exercise prior to resin injection in Feature A, a Pilot Resin Experiment has been performed in a near-drift situation in the F-tunnel, c.f. Figure 8. The site has been investigated by an array of 9 short 56mm boreholes which have been subject to different types of geological and hydraulic characterization, and a descriptive model was developed. Three injections of a two-component resin was performed using a customized injection pump. The injections were preceded by injection of water and alcohol labelled with a dye.

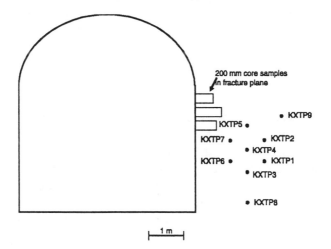

Figure 8   TRUE-1 Pilot Resin Experiment. Vertical section through the F-tunnel showing intersections of the target feature by 56 mm characterization boreholes.

The three resin batches were labelled using a combination of uranine and a dye, to enable discrimination of resin of different origin. Resin has been injected over periods of up to 7 hours under which a few litres of resin have been injected. Subsequently, the site will be excavated using 5-10 exploratory 56mm and a few 200mm cored boreholes and analysed for resin occurrence and pore space. The latter activity is ongoing.

Interpretation and modelling

Modelling using both analytical and numerical models constitutes an integral part of the scoping, design, prediction and evaluation of the tracer tests performed in TRUE-1. Tracer test interpretation methodologies applied by the SKB TRUE Project team span from preliminary evaluation using simple analytical one-dimensional advection dispersion models to models which consider the studied feature as a heterogeneous medium and use a stochastic continuum representation of transmissivity on the scale in question. Both transmissivity and velocity are thus considered as random space functions. Conditioning of these functions can me made using available data (transmissivity, head, tracer breakthrough etc.). Conservative transport and possible mass transfer are modelled based on the Lagranginan travel time approach. A separation between conservative transport and mass transfer reactions is possible in the analysis, such that an analytical treatment of mass transfer is easily combined with a numerical particle tracking scheme. The planned resin injection in Feature A and subsequent excavation and analysis of pore space will constitute a test and screening and further improvement of conceptual models of single fracture representation on a detailed scale.

Plans for the near future

The original objectives of TRUE-1 are primarily to test and develop techniques and methodologies to be employed in future experimental phases of the TRUE Programme. Retention mechanisms were thus not planned to be addressed directly/explicitly in TRUE-1. However, a series of reactive tracer tests have been added and will conclude the tracer test programme. The intention is that in future TRUE experiments several experiments will be performed, including use of a variety of tracers, flow geometries, flow rates, transport directions and distances, which will enable discrimination between active retention processes. Embedded in the comparison between the breakthrough of an ideal conservative tracer and a (weakly) sorbing tracer breakthrough for a given flow field is information on sorption characteristics of the fractures of the rock and the surface area available for sorption and matrix diffusion. Data on pore space distribution from resin injection are important for resolving the ambiguity in this inference. Moreover, it is important to enable generalization of these findings and investigate the potential for extrapolation to other features, and potentially, to other sites. The ambition should at all times be to establish a conceptual descriptive model which is valid for a variety of flow geometries, selection of tracers and flow rates. The potential of using single hole tracer tests in discriminating/verifying the action/existence of matrix diffusion from similar effects introduced by heterogeneity will also be explored.

The preliminary characterization for the TRUE Block Scale experiment will commence during the first half of 1997. The specific objectives of the TRUE BS are to; 1) increase understanding and the ability to predict tracer transport in a fracture network, and 2) assess the importance of tracer retention mechanisms (diffusion and sorption) in a fracture network, and 3) assess the link between flow and transport data as a means for predicting transport phenomena. In the block scale fracture network address, the main challenge lies in actually performing successful (controlled) tracer experiments at the desired scale. This scale is also interesting from a modelling point of view in that it constitute a possible lower bound in scale where continuum models are applicable.

Discussion and conclusions

A series of tracer experiments denoted Tracer Retention Understanding Experiments (TRUE) are underway at the SKB Äspö Hard Rock Laboratory. The objectives of the experiments are to further develop the understanding of radionuclide migration and retention in fractured rock. Integrated is a successive evaluation of concepts used in applied models as to whether they are based on realistic description of fractured rock, and to what extent adequate data can be collected in the field to inform the concepts used. In realizing the experiment a staged approach is utilized where different transport scales are addressed and the degree of complexity is successively increased. It is expected that the TRUE experiments can provide important contributions towards understanding the conditions under which radionuclide retention processes are active. Further, it is expected that the integration of results of field tracer test at different scales with laboratory experiments will provide guidance as to how to optimize field and laboratory tracer test programmes in future site characterization. Specifically, the TRUE-1 experiments show that field tracer test can be carried out successfully under the pressure and salinity conditions experienced at Äspö. It is expected that the continued experimentation at the TRUE-1 site will provide valuable contributions on transport properties and techniques to characterize the pore space of fractured rock.

Acknowledgements

The contributions from the field and analysis and modelling crews of the SKB TRUE Project team, the Äspö Hard Rock Laboratory staff and the colleagues of the Äspö Task Force are gratefully acknowledged. The work reported in this paper is funded by the Swedish Nuclear Fuel and Waste Management Company (SKB).

References

[1] Winberg, A. (ed.) et al. 1996 : Descriptive structural-hydraulic models on block and detailed scales of the TRUE-1 site. First TRUE Stage - Tracer Retention Understanding Experiments. Swedish Nuclear Fuel and Waste Management Company (SKB), Äspö Hard Rock Laboratory International Cooperation Report ICR 96-04.

[2] Dershowitz, W., Thomas, A. and Busse, R. 1996 : Discrete fracture analysis in support of the Äspö Tracer Retention Understanding Experiment (TRUE-1). Swedish Nuclear Fuel and Waste Management Company (SKB), Äspö Hard Rock Laboratory International Cooperation Report ICR 96-05.

Design, Modelling, and Current Interpretations of the H-19 and H-11 Tracer Tests at the WIPP Site

Lucy C. Meigs, Richard L. Beauheim and James T. McCord
Sandia National Laboratories, USA

Yvonne W. Tsang
E. O. Lawrence Berkeley National Laboratory, USA

Roy Haggerty
Oregon State University, USA

Abstract

Site-characterization studies at the Waste Isolation Pilot Plant (WIPP) site in southeastern New Mexico, USA identified ground-water flow in the Culebra Dolomite Member of the Rustler Formation as the most likely geologic pathway for radionuclide transport to the accessible environment in the event of a breach of the WIPP repository through inadvertent human intrusion. The Culebra is a 7-m-thick, variably fractured dolomite with massive and vuggy layers. In the 1980s, a series of tracer tests was performed in the Culebra to identify important transport mechanisms and quantify transport parameters for use in a preliminary performance assessment (PA) of the WIPP site. Comments received from numerous review and regulatory groups indicated the need to distinguish among alternative conceptual models. Based on extensive interactions with numerous review groups and outside scientists, additional testing was planned.

The results of recent tracer tests, as well as hydraulic tests, laboratory measurements, and re-examination of Cuelbra geology and stratigraphy, have led to a significant refinement of the conceptual model for transport in the Culebra. The Culebra was previously conceptualized as a medium where advection was only through fractures with diffusion into a relatively homogeneous rock matrix. The Culebra is now conceptualized as a heterogeneous medium with multiple scales of porosity. Tracer test results and geologic observations suggest that flow occurs within fractures, and to some extent within interparticle porosity and vugs connected by microfractures. Diffusion occurs within all connected porosity. Numerical simulations suggest that the data from the tracer tests cannot be simulated with heterogeneous single-porosity models; significant matrix diffusion appears to be required. The low permeability and lack of significant tracer recovery from tracers injected into the upper Culebra suggest that transport primarily occurs in the lower Culebra.

The success of the recent tracer tests in refining the conceptual model of Culebra transport was due to extensive interactions between modellers and experimenters prior to and during the tests. The test design was the direct result of interpretations of past tests and was continuously refined during the tests as additional insight was gained from evaluation and interpretation of new data. This step-wise approach was valuable for designing a robust test, but the approach could be improved by adopting an even more evolutionary strategy over a longer time frame.

Introduction

The Waste Isolation Pilot Plant (WIPP) is a proposed repository for transuranic wastes constructed in bedded Permian-age halite deposits in southeastern New Mexico, USA. Site-characterization studies at the WIPP site identified ground-water flow in the Culebra Dolomite Member of the Rustler Formation as the most likely geologic pathway for radionuclide transport to the accessible environment in the event of a breach of the WIPP repository through inadvertent human intrusion. The Culebra is a 7-m-thick, variably fractured dolomite with massive and vuggy layers. Between 1980 and 1988, tracer tests were performed in the Culebra at five locations (H-2, H-3, H-4, H-6, and H-11 hydropads) to identify important transport processes and mechanisms, and to quantify transport parameters. The information derived from these tests was used in preliminary performance assessments (PA) of the WIPP and was presented to numerous review and regulatory groups, including the International Project to Study Validation of Geosphere Transport (INTRAVAL), the U.S. National Academy of Sciences, the U.S. Environmental Protection Agency (EPA), and the New Mexico Environmental Evaluation Group (EEG). Comments received from these groups indicated the need for additional tracer tests to distinguish among possible alternative conceptual models, define transport processes and parameters more definitively, and develop a defensibly conservative simplified model of transport for PA. Through extensive interactions with these review groups and other interested scientists, a plan for tracer testing at a new test site, the H-19 hydropad, was developed. Based on the results of preliminary tests at H-19, additional testing was planned and performed at the H-11 hydropad to resolve questions associated with tests previously conducted at that location. Beauheim et al. [1] describe the rationale for additional experiments in greater detail.

The objectives of the new experiments were to test important features of the transport models used to interpret the previous tests, further evaluate transport processes in the Culebra, and provide quantitative estimates of important transport parameters. This new information was to be combined with the information derived from previous tracer tests and laboratory tests (e.g., porosity measurements, batch tests, etc.) to develop both a transport model and parameter ranges for use in PA. The initial Culebra transport model was based on interpretations of previous tracer tests performed at the H-3, H-6, and H-11 hydropads. These interpretations were based on a double-porosity model in which advective transport occurred through uniformly spaced fractures (constant matrix block size), with physical retardation provided by diffusion from the fractures into cubic matrix blocks, and directional differences in transport were caused by anisotropy in permeability. The new tests were designed to evaluate the assumptions used in this transport model. Specifically, the tests should demonstrate whether matrix diffusion occurs in the Culebra, whether an idealized fracture-matrix geometry was adequate to model test results, whether anisotropy alone could account for directional differences in transport, and the effect that layering has on transport within the Culebra.

Three elements in the design of the new tests were focused on the demonstration of matrix diffusion. First, single-well injection-withdrawal (SWIW) tracer tests were planned that were expected to distinguish clearly between the effects of matrix diffusion and heterogeneity in permeability. Second, tracer injections were to be repeated in convergent-flow tests performed at different pumping rates to show the effects of advective residence time on diffusion. Third, two different conservative tracers having different free-water diffusion coefficients were to be injected together in a convergent-flow tracer test to show the effects of different amounts of diffusion. See Beauheim et al. [1] for a discussion of pre-test calculations for these three design elements.

Results of Recent Tests

Testing began at the H-19 hydropad in June 1995 with SWIW and preliminary convergent-flow tests after wells H-19b0, b2, b3, and b4 had been completed (Figure 1). The purpose of the preliminary convergent-flow test was to provide site-specific data for the determination of locations of three additional wells, and to provide data for test design refinements and model testing. The breakthrough curves obtained from the convergent-flow test showed transport to be slower at H-19 than at other sites tested. As a result, additional wells were drilled closer to H-19b0 than had been projected without site-specific data. Numerical simulation indicated that the preliminary convergent-flow test data could not be matched with the range of parameters and conceptual model used to match tracer tests conducted previously at other sites. To match the preliminary test data approximately, the advective porosity had to be greater than 0.03, which is larger than typical fracture porosities. This result raised questions concerning the validity of conceptualizing the Culebra as a medium where advection was only through fractures with diffusion into a relatively homogeneous rock matrix. The results of the preliminary test also led to refinements in the design

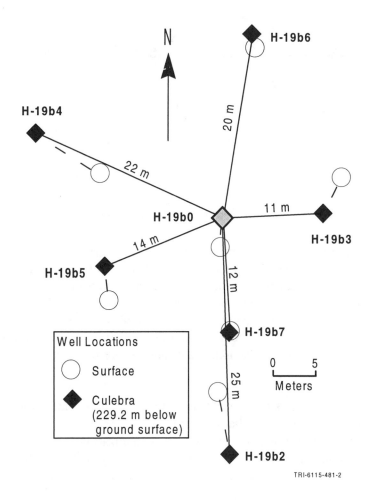

Figure 1. Well locations at the H-19 hydropad.

of tests performed after all seven wells had been completed, as well as to the decision to perform additional tests at the previously tested H-11 hydropad (Figure 2) where significantly faster transport had been observed. Conducting a simpler and shorter preliminary tracer test prior to the subsequent complicated and lengthy tracer test provided valuable information for refinement of test design, refinement of conceptual models, and provided a 'dry run' for testing equipment and procedures used in the field and laboratory.

Starting in October 1995, hydraulic tests were performed in the seven wells at the H-19 hydropad. These tests revealed that the permeability of the upper portion of the Culebra was significantly lower than that of the lower Culebra at this site [1]. The information obtained from the hydraulic tests was used to refine the tracer-test design. Tracer testing resumed at H-19 in December 1995, with a SWIW test of the lower Culebra followed by convergent-flow testing with tracer injections in six wells. At three wells, tracers were injected into both the entire thickness of the Culebra and into upper and lower portions of the Culebra. Some of the flow paths were tested again beginning in February 1996 after the pumping rate had been reduced. Testing at H-11 began in February 1996 and consisted of a SWIW test over the entire thickness of Culebra, followed by convergent-flow tests involving two injection wells and two pumping rates.

The breakthrough curves for the six pathways tested with the multi-well convergent-flow test at H-19 show significant variations (Figure 3). For example, the fastest peak arrival time is not

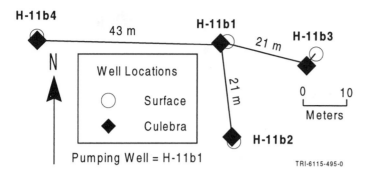

Figure 2. Well locations at the H-11 hydropad.

from the shortest travel distance and the slowest peak arrival time is not from the longest travel distance, indicating that transport is through a heterogeneous and/or anisotropic medium. The peak heights of breakthrough curves at H-11 and H-19 for given pathways (e.g., H-11b3 to H-11b1) for the two different pumping rates used are the same within measurement error (Figure 4). Tracers injected into the upper portion of the Culebra at H-19 were only detected at the pumping well in very low concentrations. The iodide data from the three multi-well injections where iodide and a benzoic acid were injected simultaneously were difficult to obtain and less reliable due to the high salinity of the Culebra brine. Despite these difficulties, the normalized concentration breakthrough curves, at all three locations, suggest that the peak height of the iodide was less than the peak height of the benzoic acid, as would be expected if matrix diffusion is occurring because iodide has a larger free-water diffusion coefficient than the benzoic acid. The difference between the iodide data and the benzoic acid data is greater at H-11 (Figure 4) than at H-19 (Figure 3).

The data from the breakthrough curves and subsequent modelling results (e.g. the large advective porosities that appear to be necessary to model the tracer breakthrough data from H-19) motivated refinement of the conceptualization of transport in the Culebra. Through careful reexamination of the geology and stratigraphy of the Culebra, a more thorough conceptualization has

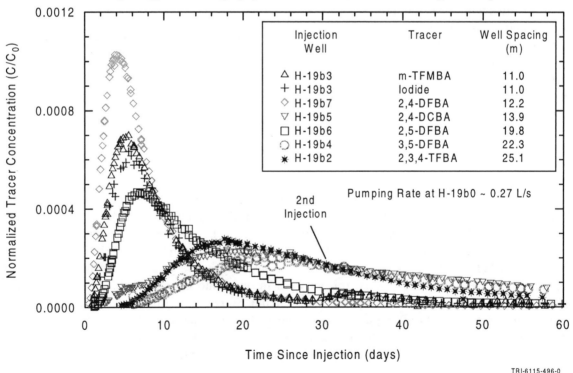

Figure 3. Observed tracer data (e.g., 2,3,4-trifluorobenzoic acid) for the H-19 multi-well convergent-flow test at the high pumping rate.

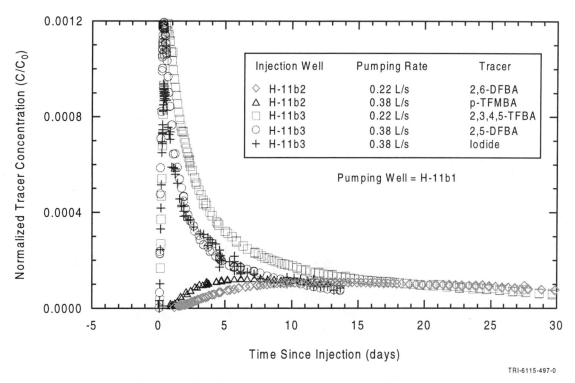

TRI-6115-497-0

Figure 4. Observed tracer data for the H-11 multi-well convergent-flow test at the high and low pumping rates.

been developed of the important processes that control transport. The Culebra has non-uniform properties both horizontally and vertically as was demonstrated by both hydraulic and tracer tests. The low permeability and lack of significant tracer recovery from injections into the upper portion of the Culebra suggest that transport primarily occurs in the lower Culebra. Multiple scales of porosity exist within the Culebra, including fractures ranging from microscale to large, vuggy zones, intercrystalline, interparticle and intraparticle porosity (Figure 5). Laboratory (core plug) measurements of porosity of the Culebra yield values between 0.03 to 0.30 (median of 0.16), which indicates that there is significant porosity for advection and diffusion. Tracer test results and geological observations suggest flow can occur within fractures, and to some extent within interparticle porosity and vugs where they are connected by fractures. Diffusion occurs within all connected porosity and will be the dominant transport mechanism in relatively low permeability portions of the formation. The variation in peak arrival times in tracer-breakthrough curves between the H-11 and the H-19 hydropads suggests that the types of porosity contributing to rapid advective transport may vary spatially.

Numerical Simulations of Tracer-Test Data

Interpretations of the tracer-test data completed to date have relied on both homogeneous and heterogeneous single- and double-porosity continuum models. In the double-porosity models, the Culebra is conceptualized as two continua: the advective porosity (fractures, vugs, and possibly interparticle porosity) where flow is the dominant process and the diffusive porosity (all other connected porosity) where diffusion is the dominant process. Spatial variations in advective transport are represented in numerical simulations of tracer-test data with unconditioned, spatially correlated, random hydraulic conductivity fields. Diffusive transport was simulated with double-porosity models using both a single rate of diffusion (which conceptualizes diffusion as a homogeneous process) and multiple rates of diffusion (which conceptualizes the diffusive porosity as heterogeneous). The multi-

161

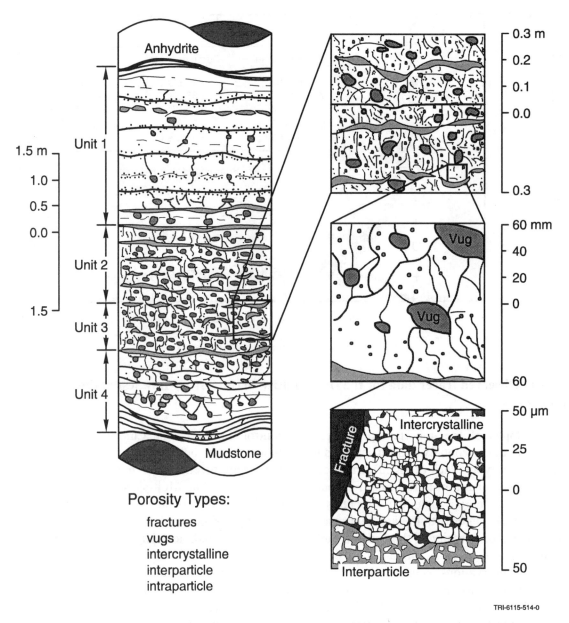

Porosity Types:

fractures
vugs
intercrystalline
interparticle
intraparticle

TRI-6115-514-0

Figure 5. Schematic illustration showing multiple scales of Culebra porosity.

rate model employed [2] represents the heterogeneities in block size (surface area for diffusion and diffusion distance) and the tortuous nature of the pore structure with a distribution of diffusion rates (Figure 6).

Interpretations completed to date have shown that the SWIW test data from both the H-11 and H-19 hydropads cannot be explained by heterogeneity in hydraulic conductivity alone. Simulations of cumulative mass recovery during the withdrawal phase of the tracer test with homogeneous and heterogeneous single-porosity models suggest that mass recovery for these conceptual models should be very rapid. Figure 7 shows the mass recovery from the H-11 SWIW test compared to heterogeneous single- and double-porosity simulations. The H-11 (and H-19) data show a slow mass recovery as would be anticipated if some form of diffusional process were controlling mass

162

recovery. Simulations further suggest that the SWIW data cannot be adequately explained using a single-rate double-porosity model with homogeneous or heterogeneous permeability fields. With a single-rate double-porosity model, the simulated tracer recovery concentration during a SWIW test will asymptotically approach a constant slope of -3/2 on a log-log plot of normalized tracer concentration vs. elapsed time (Figure 8). If matrix blocks are small and the solute diffuses to the centre of the block, the diffusion rate will decrease and the late-time slope will steepen. The late-time slope of the data from the H-11 and two H-19 SWIW tests was approximately -5/2 rather than -3/2. An excellent fit to the data can be obtained using the multi-rate diffusion model that incorporates a statistical distribution of mass transfer rates. Figure 9 shows fits to both the H-11 and H-19 SWIW test data with the multi-rate model. Figure 10 shows the distributions of diffusion rates that were used to match the data in Figure 9. For these distributions, if tortuosity is assumed to be constant at 0.1, the mean block sizes would be 0.03 m and 0.006 m for the H-11 and H-19 simulations, respectively. These diffusion rate distributions cannot be considered to represent unique solutions because equally good fits could be obtained using different values of advective and diffusive porosity that would produce somewhat different distributions.

The multi-well convergent-flow data also appear to exhibit evidence of heterogeneity and matrix diffusion. Preliminary attempts to match the convergent-flow data with a single anisotropy were unsuccessful. Heterogeneity, as opposed to simple anisotropy, appears to be required to capture the variations in the six tracer-test breakthrough curves. Numerical simulations of tracer-test data with heterogeneous single and double-porosity models suggested that the data cannot be adequately modelled without matrix diffusion (Figure 11). Heterogeneous single-porosity models cannot simultaneously match the magnitude of the peak

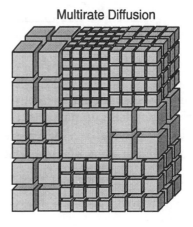

Single-Rate Diffusion

Multirate Diffusion

TRI-6115-513-0

Figure 6.   Schematic of double-porosity models. Single-rate diffusion models have a constant matrix block size (i.e., surface area for diffusion and diffusion distance) and constant tortuosity (tortuous nature of 'matrix' pores). Multiple rate diffusion models have a distribution of diffusion rates attributed to variations in matrix block size and tortuosity.

concentration and peak arrival time. Heterogeneous single-rate double-porosity models can match individual breakthrough curves quite well. However, when the parameters used to match data from the high pumping rate (0.27 L/s) are used to simulate the breakthrough curve for the low pumping rate (0.16 L/s), the match is not satisfactory (Figure 12). With a single-rate double-porosity model with a given pumping rate, as matrix block size decreases, peak heights gradually first decrease and then increase (Figure 13). The increase in peak height occurs as a result of solute diffusion reaching the centre of the matrix block, which causes a decrease in the concentration gradient and, therefore, a decrease in the diffusion rate. As a result, peak heights will not always be lower for lower pumping rates. Simulations have shown that a model with multiple rates of diffusion can produce peak heights that are similar for the two pumping rates used for the H-19 testing (Figure 14).

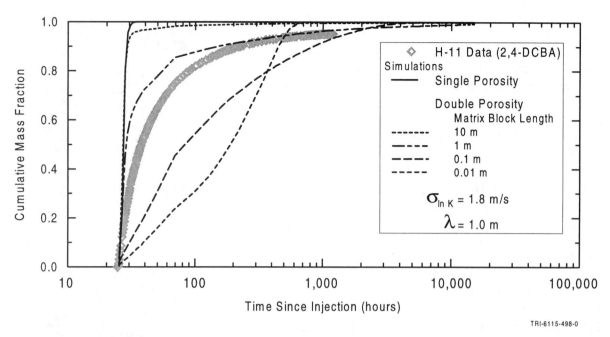

Figure 7. Simulated mass recovery curves for H-11 SWIW test with single-porosity and double-porosity models compared to observed data. For the simulations, the heterogeneous field of hydraulic conductivity had a standard deviation (σ) in natural log space of 1.8 m/s and an exponential model with a correlation length (λ) of 1.0 m.

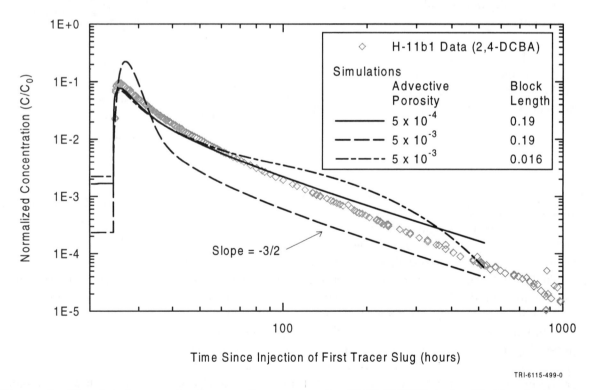

Figure 8. Simulated concentration vs. time curves for H-11 SWIW test with the single-rate double-porosity model compared to observed data.

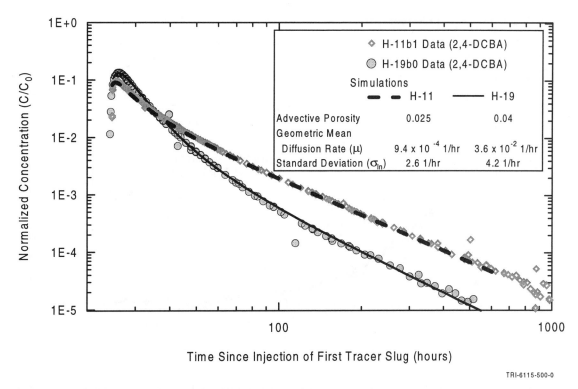

Figure 9. Simulated concentration vs. time curves for H-11 and H-19 SWIW tests with the multi-rate double-porosity model compared to observed data.

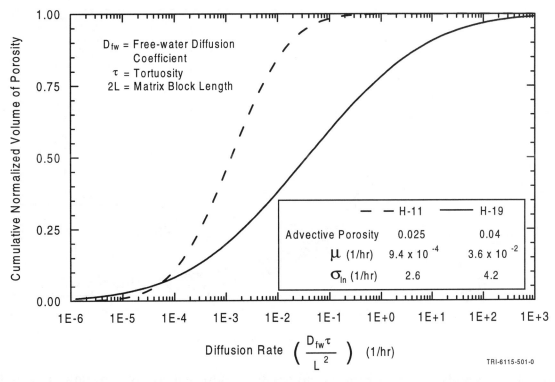

Figure 10. Diffusion rate coefficient cumulative distribution function for H-11 and H-19 SWIW test simulations shown in Figure 9.

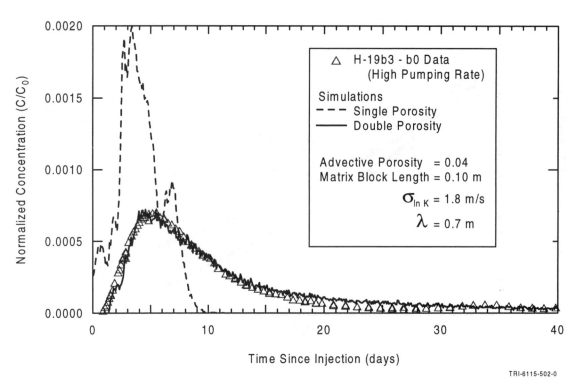

TRI-6115-502-0

Figure 11.   Simulated breakthrough curves for multi-well convergent-flow test data, H-19b3 – b0 path, using heterogeneous, single- and double-porosity models.

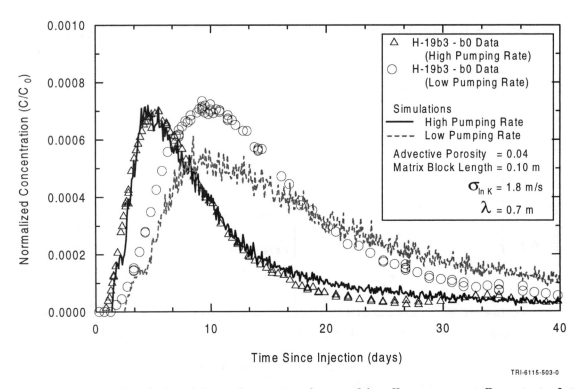

TRI-6115-503-0

Figure 12.   Simulated breakthrough curves for multi-well convergent-flow test data, H-19b3 – b0 path, using heterogeneous, single-rate double-porosity model for two different pumping rates.

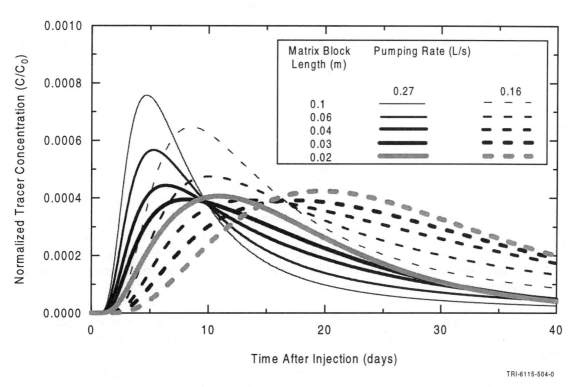

Figure 13. Effect of pumping rate and block size on peak height and breakthrough time.

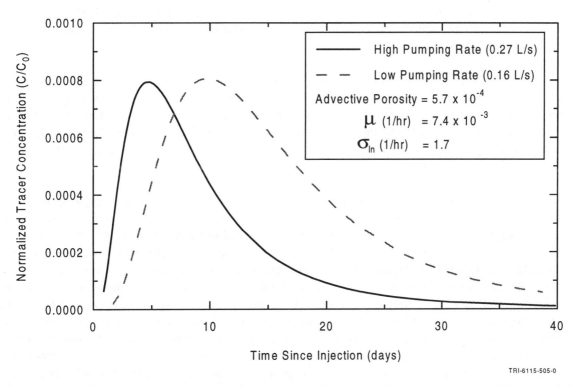

Figure 14. Simulations of two different pumping rates with multi-rate double-porosity model.

167

The data collected from recent tracer tests and numerical simulations to date have led to a significant refinement of the conceptual model of the Culebra. Numerical simulations of the multi-well convergent-flow and SWIW tests suggest that the data cannot be simulated with heterogeneous single-porosity models; significant matrix diffusion appears to be required. Numerical simulations of the multi-well tracer tests suggest that advective transport is not limited to fractures and that there is significant variation in advective porosity across the WIPP site.

Discussion

For the larger-scale simulations conducted for PA, spatial variability in advective transport is represented by heterogeneous transmissivity fields that have been conditioned on available point transmissivity data and transient pressure data from large scale pumping tests. In the PA calculations, the lower permeability of the upper portion of the Culebra has been approximated by eliminating this portion of the Culebra from the transport model. Numerical simulations of tracer test data were used to bound transport parameters for use in PA models. Possible spatial variability in transport properties (diffusion and sorption rates) has not been treated explicitly in the PA model to date. Attempts have been made to take this variability into account by providing PA with conservative ranges of values for transport parameters (i.e., values that could lead to greater releases than expected).

Tracer tests appear to be well suited for providing insight into the important processes that are operative at a given site and for testing conceptual models. For example, the mass-recovery curve from the SWIW test appears to be well suited to evaluating the importance of diffusive processes. Complex tests are valuable for evaluating conceptual models. A clear conceptual model that incorporates the tracer test results with what is understood about the geology and stratigraphy at the PA scale is important if tracer test results are to be used to provide transport properties over a much larger region. Tracer tests will always have limitations: they cannot directly test the materials for the length and time scales of interest for PA calculations. Thus, one must evaluate whether alternative conceptual models can explain the data, particularly if different conceptual models would lead to different results at the PA scale.

The success of the recent tracer tests in refining the conceptual model of Culebra transport were the result of extensive interaction between the modellers and the experimenters prior to and during the tests. The details of the design were continuously refined during the tests as additional insight was gained from evaluation and interpretation of new data. The iterative evaluation of test data was especially important for such test-design features as injection into different layers of the Culebra and the selection of different pumping rates.

The testing approach used could be improved by adopting a more evolutionary strategy over a longer time frame. The step-wise strategy that was used, which included conducting a preliminary tracer test prior to completion of all wells, was very valuable for developing a robust tracer-test design. Ideally, field testing would begin after only one or a very few wells were installed. The early tests would be completely interpreted before plans for additional wells and tests were finalized. In parallel with field tests, detailed laboratory tests would be performed. The laboratory tests would provide a controlled environment for testing rock diffusion and sorption properties. The later field tests and well locations would be designed to address questions/ambiguities raised by the early tests in both the field and lab. New models/hypotheses raised by the new tests would be evaluated through tests at other locations. Design of new tests would be well integrated with ongoing laboratory programs. Sorbing tracers would be added to the programme at a late stage to evaluate conclusions reached from laboratory testing. Specific experimental enhancements would include increasing the contrasts in pumping rates and possibly tracer diffusion coefficients, improved control over rates/pressures of tracer injection to allow interpretation of hydraulic responses, and development of a reliable means of determining the downhole tracer source term. Laboratory testing and modelling would be used to provide confidence that the downhole source term could actually be measured under the anticipated test conditions before investing in a field-scale tool and field testing.

Summary

Extensive interactions between experimenters and modellers resulted in a robust test design, which lead to significant refinement in the conceptual model for transport in the Culebra. Based on hydraulic tests, tracer tests, and geologic descriptions, the Culebra is now conceptualized as a heterogeneous medium with multiple scales of porosity. The low permeability and lack of significant tracer recovery from tracers injected into the upper Culebra suggest that transport primarily occurs in the lower Culebra. The tracer test data provide evidence that transport is not limited to fractures. Numerical simulations of the multi-well convergent-flow tracer test suggest that advective transport is not limited to fractures. Advection also appears to occur to some extent through interparticle porosity and vugs connected by microfractures. Numerical simulations of the multi-well and SWIW tests suggest that the data cannot be simulated with heterogeneous single-porosity models; significant matrix diffusion appears to be required. Numerical simulations of the tracer test data with this refined conceptual model were used to bound transport parameters for use in PA models.

Acknowledgments

This work was supported by the US Department of Energy (US DOE) under contract DE-AC04-94AL85000. Sean McKenna of Sandia National Laboratories did the simulations in Figures 11 and 12. Joanna Ogintz and Toya Jones of INTERA did the simulations in Figures 8 and 13. Robert Holt provided the geologic information for creation of Figure 5. Numerous members of the Sandia National Laboratories and INTERA staff helped to make these tracer tests possible.

References

[1]   Beauheim, R.L., Meigs, L.C., and Davies, P.B., Rationale for the H-19 and H-11 Tracer Tests at the WIPP Site, 1996, [this volume].

[2]   Haggerty, R., and Gorelick, S.M., Multiple-rate mass transfer for modeling diffusion and surface reactions in a media with pore-scale heterogeneity, *Water Resources Research*, vol. 31, no. 10, p. 2383–2400, 1995.

[3]   Hadermann, J., Heer W., The Grimsel (Switzerland) migration experiment: integrating field experiments, laboratory investigations and modelling, *Journal of Contaminant Hydrology*, vol. 21, p. 87-100, 1996.

Design, performance and interpretation of tracers tests at El Berrocal site (Spain)

J. Guimerà (1); M. García-Gutiérrez (2); A. Yllera (2); J. Carrera (1); A. Hernández-Benítez (2) and M. Saaltink (1)

(1) Universitat Politècnica de Catalunya, Barcelona (Spain)
(2) Centro de Investigaciones Energéticas Medioambientales y Tecnológicas, Madrid, Spain

Abstract

Tracer tests at El Berrocal site were aimed to hydrological characterization, instrumentation development and data base generation for modeling purposes. They were placed at two borehole arrays, involving different degrees of fractured granite. Tracers were exclusively non-sorbing although great differences were observed among them during the recovery periods. Newly developed instrumentation proved to be robust and reliable working with downhole electronics. Interpretation of the tests included several conceptual models. Those that resulted in more realistic prediction capabilities took into account the heterogeneity of the system, matrix diffusion and the hydrogeological context at local scale. Simple analytical models proved to be helpful but with limited prediction capabilities. Finally, including in the model features and processes, such as the bore presence or the transient of levels during the tracer injection, which are often neglected in normal permeability media, was of utmost importance

1. Introduction. The Concept

El Berrocal project was an international effort to study radionuclide migration under natural conditions in a granitic environment. El Berrocal site was not considered as a potential repository site. The particular interest of tracer tests at El Berrocal was, on one hand, to enhance hydrogeological characterisation and, on the other, to build know-how on performance and instrumentation development. In the frame of R&D programmes of ENRESA and the CEC, El Berrocal project brought together research on flow and transport in heterogeneous domains, laboratory migration experiments and a vaste amount of field work. A close link of theoricists, modelers, lab and field experimentalists summed-up the background experience to design, perform and interpret a series of tracer tests which were reasonably satisfactory. Once the design of the experiment reached a preliminary stage, it was submitted for discussion to an international group of experts on tracer tests. Their recomendations were of great help and were incorporated into the planning and performance of the experiments. As a result, block-scale characterization benefited from a multidisciplinary team that had been working in separated areas in the near past.

As most of the tracer tests, they were performed at the most conductive fractures between boreholes. Therefore, results were biased towards relatively high permeability zones. That would cast serious doubts on the validity of the results, in terms of reducing the uncertainy due to spatial variability. To overcome this limitation, interpretation accounted explicitly for either conductive fractures and rock matrix. Besides, rock mass double porosity models were used to simulate processes likely to occur such as matrix diffusion.

Indeed, in spite of conservative tracers were used, neither of them behaved the same to each other. All of them displayed similar first arrival times; but peak arrival times were different, corresponding the tracers of biggest molecule sizes to the latest times. Differences in tailing were significant. Therefore, we conclude that results are conditioned by matrix diffusion and that such process may play an important role even in weakly reactive tracers. This contribution is important in terms of long term predictions of contaminant transport. Models orientated towards risk management-PA, may benefit from these results, even though the whole project was not intended for that. Regional and local geological, geochemical and hydrogeological settings are out of the scope of this paper. The reader is refered to [1] for general readings of the project.

2. Design

The tests involved two arrays of boreholes (Figure 1): S11-S12 (vertical) and S2-S13-S15 (inclined). Both intersected several fracture sets, at depths down to 70 m. Adequately controlling the mass flux in and out of the system is difficult on tracer tests performed in low permeability media, where water fluxes are small and residence times at boreholes may be high. In order to overcome this problem, we designed a system to both take samples and obtain continuous records of selected tracer concentrations, both at the injection and extraction points. The system could also monitor (1) pressure and temperature at the isolated sections, (2) tracer injection and the homogeneization recirculation flow rates at injection intervals, (3) recirculation and extraction rates at the output, and (4) pressures at recirculation systems. Samples were taken automatically at variable time intervals and were checked manually. Further details may be found at [2].

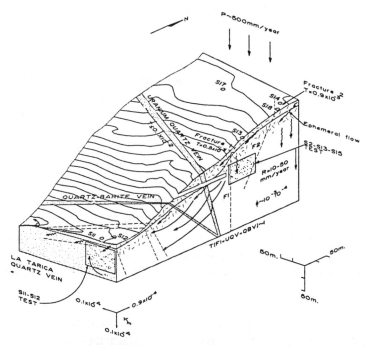

Figure 1. Block diagram of El Berrocal site involving major elements of the hydrogeological conceptual model (after Guimerà et al. 1996). Boreholes involved in tracer tests were 11-12 (vertical) and 2-13-15 (inclined).

Tests methodology allowed us not only to identify preferential flow paths, but also to evaluate how would they be affected by the convergent flow test. Dilution tests, performed both with and without pumping, were of great help in this context.

Tracer selection considered water isotopes, salts, dyes and compounds such as Gd-DTPA and Re as ReO_4^-. An attempt was made using selenium Se(VI) which was not successful (recovery was not achieved where other tracers had been recovered previously). Retardation was characterized in column experiments, which showed retardation to be small. Yet, we think that tracer sorption on equipment and wellbore walls was not negligible, and much more important that in lab conditions. This effect may have been due to differences in tracer concentration at laboratory and field experiments.

Low permeability causes flow rates to be small, which leads to long tests duration (from few weeks to months). As a result, the experiments can be affected both by variations on natural conditions (rainfall, recharge and changes in piezometric levels) and by instrumentation failures (the most typical being changes in pumping rate). The total duration of the test can be shortened by reducing injection time, which can be achieved by pulse injection, which we aimed at by following the procedure described below.

Before injecting the tracers, a recirculation flow was established at both the injection and pumping points, with a flow rate 4 to 10 times higher than the pumping rate. Then, one or several tracers were injected. When the concentration at the injection section was homogeneized, the tracer was forced to flow into the medium by injecting a known volume of clean water. This injection caused a small perturbation in the water flow that lasted a few hours, but this time is small when compared to the total duration of the experiment (more than 30 days). It is possible that, due to this injection, part of the tracer was carried out of the range of the pumping borehole, which eventually would explain that the tracer was not fully recovered. The main advantage of this type of injection is that over 90% of the tracer gets into the formation before the end of the first day.

Tracers were injected at two isolated sections of borehole 11 and recovered at one pumping section of borehole 12 at the 11-12 test. The 2-13-15 tests consisted of one flushed injection at one section of borehole 13 and three decay injections at borehole 13 (2) and 15 (1). Table 1 summarizes the performances. [2] and [3] report further details of the performance of the tests.

3. Results and Interpretation

This section summarizes the results achieved in the test at 11-12 and the tests at 2-13-15 boreholes. Accordingly, modeling of each of them are presented. Results are displayed as breakthrough curves which are shown in te figures of model runs.

3.1 Test at 11-12

3.1.1 Analytical interpretation

We modeled the three breakthrough curves independently by means of the computer code TRAZADOR [4], which estimates the parameters of the conservative transport equation. That lead to a best fitting of the concentration data using the maximum likelihood method. The models chosen for interpretation were radial advection-dispersion and radial advection-dispersion with matrix diffusion.

Interval	Action	Depth (m)	Distance to recovery (m)	Dilution volume (l)	Flushing volume (l)	Tracers	Rec (%)	10% peak arrival (d)	Peak arrival (d)	Peak C
TEST 11-12										
12	pumping 2 l/min	49-27	-	129	-	-	-	-	-	-
11-1	Injection	64-48	22.0	118	385	Uranine 5 g	62.5	3.0	7.5	60 ppb
						Iodide 3.5 g	61.3	3.0	7.5	120 ppb
11-2	Injection	47-25	14.5	140	300	Eosine 5 g	41.2	1.7	7.5	60 ppb
TEST 2-13-15										
2	Pumping 0.09 l/min	20.79-30.10	-	53.55	-	-	-	-	-	-
13-4	Injection	31.09-37.08	19.6	37.52	300	D$_2$O 12 l	56.3	0.5	2.2	rel. unit
					300	Uranine 15 g	40.8	0.5	4.2	255 ppb
					300	Gd-DTPA 1.1 g	43.5	0.5	2.0	435 ppb
					0	Phloxine 12.4 g	30.1	3.5	14.1	0.90 ppr
					0	Re 0.7 g	34.5	3.5	14.0	0.06 ppr
13-2	Injection	66.66-73.70	22.2	49.04	0	B.Sulpha. 13.1 g	0.0			
					0	Iodide 11.4	0.0			
15-2	Injection	35.64-42.63	24.6	49.38	0	Eosine-Y 13.5 g	0.0			

Table 1. Summary of the performances and results of tracer tests at El Berrocal.

Implementation of the conceptual model

We assumed the flow field to be radial in spite of the geometry of the test and the fracture distribution. Pumping rate applied at borehole 12 affected the isolated intervals at borehole 11 in a different way. We had to approximate how much water was reaching each of the intervals and to distribute the pumping flowrate according to measured drawdowns and distances between intervals. As a result, flowrates were 0.625 ml/min for eosine and 1.375 ml/min for uranine and iodide (see Table 2).

Tracer	Model	L(m)	Q(l/min)	Rec (%)	ϕb(m)	α_L(m)	ϕ'_m(-)	D_m(-)
Uranine	AD	22.0	1.375	62.54	0.0187	3.400	-	-
Iodide	AD	22.0	1.375	61.30	0.0224	4.000	-	-
Eosine	AD	14.5	0.625	41.16	0.0201	4.400	-	-
Uranine	MD	-		-	0.0088	0.565	2.000	0.5479
Iodide	MD	-		-	0.0081	0.570	2.000	0.7200
Eosine	MD	-		-	0.0078	0.904	2.000	0.6039

AD= Advection-dispersion MD= Matrix-difusion ϕb= thickness porosity α_L = longitudinal dispersivity ϕ'_m = dimensionless matrix porosity D_m = dimensionles molecular difusion

Table 2. Parameter values returned by different conceptual models for the test at 11-12 (analytical model).

Since some stops occurred during the experiment, data most influenced by such failures were not used during the calibration. Instead, they are represented by different symbols in Figure 2.

Results for radial advection-dispersion

Calibration of eosine resulted in a reasonably well fitted curve, especially the rising part of the curve (Figure 2). The model returns relatively high dispersivity when reproducing the tail of the curve as well as thickness porosity (α_L = 4.4 m and ϕb = 0.02 m, Table 2). As a result, data belonging to the first two days are not properly fitted. This is probably the counterpart of maintaining the tail of the curve reasonably well reproduced. Iodide and uranine calibration returns similar parameter values: thickness porosity around 0.02 m and dispersivities 4.0 m and 3.4 m respectively. Given the quality of data, fittings are acceptable. The rising part of the uranine breakthough is specially well fitted, compared to eosine, as well as iodide. However, the model fails when reproducing the uranine tail and fits iodide rather succesfully. Differences in dispersivity are of 0.6 m, the one for uranine being smaller. Therefore, we can infer that processes affect selectively iodide and uranine in spite of their conservative nature.

Results from advection-dispersion and matrix diffusion

Eosine calibration, largely benefits from considering matrix diffusion, especially tail fitting. However, improving the fitting of the tail results in a worse reproduction of the rising part of the curve. Thickness porosity and dispersivity are being reduced by a factor of 4 (Table 2), which results in a different slope of the first part of the curve (Figure 2).

As in the previous case, the fitting of the uranine tail is improved with the inclusion of matrix diffusion. Thickness porosity is reduced by a factor of about 3, while dispersivity decreases by a factor of 8. We used the parameters returned by the uranine model to calibrate iodide breakthrough. In this case, the model does not lead to a significant improvement in the fitting, which was remarkably good with advection-dispersion, given the noisy nature of the data. Matrix diffusion fitting provides lower computed values than the measured data of the tail -as for uranine-, while those for advection-dispersion were consistently higher. Parameters returned for eosine -matrix diffusion- showed thickness porosity values similar to iodide and uranine, but longitudinal dispersivity is two-fold. Molecular diffusion also displays higher values than uranine and lower than iodide.

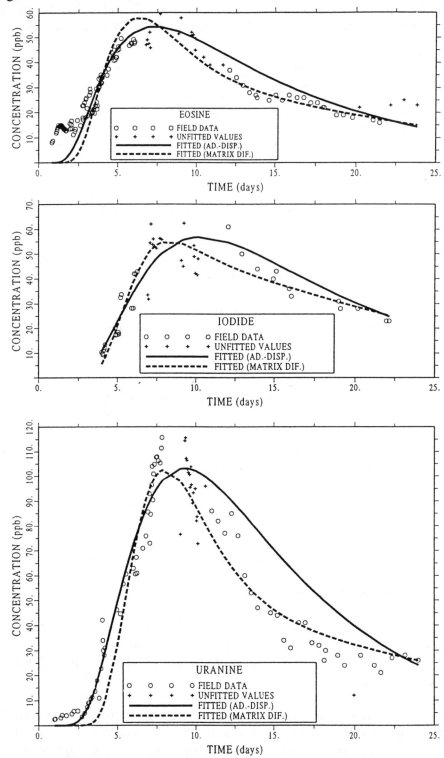

Figure 2. Results from test at 11-12 boreholes (measured data and analytical models calibration).

In summary, accounting for matrix diffusion leads to qualitative improvements on the fit and, consistently, to more reasonable parameters than advection-dispersion model. In part, this is a result of increasing the number of degrees of freedom for the model to come close to the data. Yet, the fact that the parameters are so consistent lends credibility to results. It should be kept in mind, however, that tailing can be atributed to factors other than matrix diffusion. In fact, we feel that it is caused by spatial heterogeneity ([5]) and by residual mass at the injection source. The noisy nature of data does not allow further analysis.

3.1.2 Numerical interpretation of the results

The three breakthroughs were modeled numerically as well, by using the computer code TRANSIN-III [6]. The steps we followed can be summarized as conceptual model definition, numerical model construction and automatic parameter calibration.

Conceptual model

The tracer test was started after a relative stabilization of drawdowns: we consider the flow field reached steady-state after two days of pumping. Yet, while the flow field is steady, the transport problem is transient.

Conceptual model for flow is discussed by [7]. Groundwater flow takes place through preferential flow paths, which were explicitly modeled. They were derived from structural evidence and hydraulic testing ([8]). The most important flow paths were La Tarica vein, the phreatic surface and fractures F11.4 and F11.7. Other features included in the analysis were prescribed recharge, La Tarica stream, regional flow (upwards) and pumping at S12. Water injection during tracer flushing periods was neglected due to its small volume (885 l) compared to the extraction flowrate (2.88 m^3/day). We considered that transport was affected only by advection and dispersion. Numerical interpretation of these tests considered advective-dispersive transport only. Neither retardation processes such as adsorption nor matrix diffusion were simulated.

Numerical model

[8] discuss the flow model which reproduced the steady-state conditions used to solve the transport problem. Here, we discuss only those features relevant to transport.

Molecular diffusivity is assumed to be constant throughout the whole domain, while one zone of α_L and α_T exists for each hydraulic conductivity zone. F11.7 and F11.4 are grouped into the same dispersivity zone. Dispersivity values are modified according to element sizes in order to ensure numerical stability throughout the whole domain.

Rock formation, La Tarica vein and the phreatic surface layer are assumed to have the same porosity (zone 1). Zone 2 of porosity includes fractures F11.1 and F11.7. Injection intervals are represented by zone number 3, while the remaining borehole intervals belong to zone 4. Porosity at zone 3 is fixed and equal to a very small value -10^{-5}- to simulate a rapid mixing of tracer along borehole interval and rapid movement from the hole to the formation. However, zone 4 displays fixed values equal to 1 to simulate the hole.

Calibration

We calibrated flow and transport models separately. Table 3 shows the parameters returned after flow model. Calibration of the three breakthroughs led to estimated dispersivities and porosities from both formation and fractures. Because of the large number of unknown parameters, some of these parameter values were fixed in some of the runs, according to the results of previous calibrations. This process resulted to be dilatory, yet automatic calibraton of all parameters resulted in highly uncertain values. Table 4 shows the results returned by the transport models and Figure 3 shows the corresponding fits.

Parameter	Estimated value
K in N-S direction (m/d)	0.25e-3
K in vertical direction (m/d)	0.94e-4
K in E-W direction (m/d)	0.097
T of the phreatic surface (m^2/d)	0.609
T of La Tarica vein (m^2/d)	0.022
T of fracture 11.7 (m^2/d)	0.033
T of fracture 11.4 (m^2/d)	0.428
Recharge to the phreatic surface (mm)	222.3
Constant level at La Tarica stream (m)	1.01
Constant level at bottom (m)	0.9903
Constant level at the interception of the stream with fracture 11.7 (m)	0.9833
Constant level at the interception of the stream with fracture 11.7(m)	0.8455
Leakage coefficient (1/m)	100.
Pumping rate (m^3/d)	-2.88

Table 3. Flow parameters returned by the numerical model of test at 11-12.

Parameter	Eosine	Iodide	Uranine
Longitudinal dispersivity of the rock (m)	5.24	1.75	3.32
Trasversal dispersivity of the rock (m)	0.26	0.20	0.20
Porosity of the rock (-)	0.23e-3	0.50e-3	0.35e-3
Longitudinal dispersivity of the fracture (m)	0.70	0.70	0.70
Trasversal dispersivity of the fracture (m)	0.40	0.40	0.40
Thickness porosity of the fracture (m)	0.005	0.005	0.005
Injected mass (g)	5.022	3.35	5.05
Measured recovery (%)	35	53	55
Calculated recovery (%)	35	65	65

Table 4. Transport parameters returned by the numerical model of test at 11-12.

We consider the fittings acceptable, although measured and fitted peak concentrations do not match exactly (especially true for uranine). Tailing makes the three fitting curves to be sligthly different from the measured ones. Such problems could be minimized or overcome by calibrating together flow and transport parameters, which was not the case, due to restrictions in the required computational time. More complicated models, accounting for transient flow effects or including matrix diffusion terms, could benefit the final fitting, as it was the case for the analytical interpretation. However, as mentioned earlier, we faced severe problems of uncertainity due to the large number of calibrated parameters. We feel that, even though more sophisticated models could improve the fitting, the resulting models would be less robust. Estimated values of fracture porosity (aperture) are constant for all tests. However, some

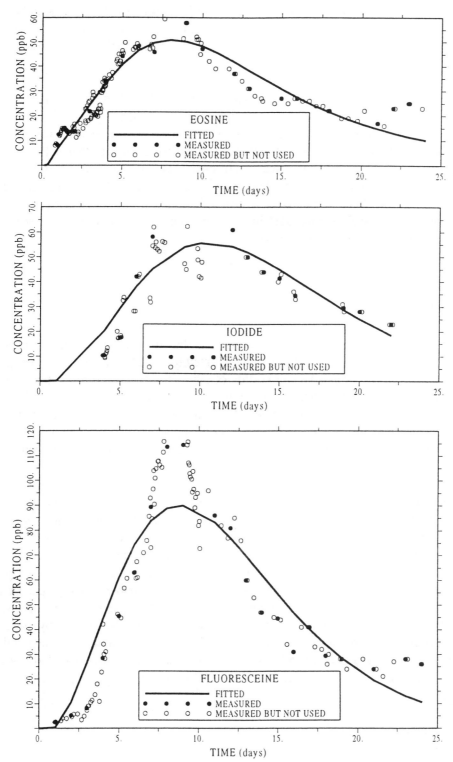

Figure 3. Fitting of the breakthroughs of test at boreholes 11-12 by automatic calibration of the 3D model.

of the dispersivity values obtained present high uncertainty. Recovered mass by the models is comparable to that obtained during the experiment. The higher recovery obtained for the calculated breakthroughs, may be due to the fact that flushing periods are neglected. Tracer flushing affects the injections in two major ways. First, part of the mass may be expelled out of the pumping area, thus reducing the recovery. Second, it can increase tailing effects, because part of the mass is forced into the rock mass. In natural or convergent flow conditions, the tracer would mainly move along major paths such as fractures.

Regarding the comparison with respect to the parameter values obtained by the analytical models, one must say that conceptual models (flow and transport) are too different to make any resembling meaningful. However, it is worth noting that the fittings obtained by the matrix diffusion model are comparable to the heterogeneous advective-dispersive model. It would appear that radial models do not realistically apply for this case, unless we make use of processes, such as matrix diffusion. Therefore, the role of matrix diffusion may not be as important as stated by the homogeneous analysis. We demonstrate that part of the effects attributed to matrix diffusion, can be explained by the heterogeneity of the media (that is, fracture flow embedded into less pervious formation). We will see in the next section that both contributions cannot be separated.

3.2 Numerical Interpretation of tests at 2-13-15

Conceptual Model

The conceptual model for the numerical interpretation was based on the hydraulic conceptual model described by [7], where complete natural 3D flow is acounted for, and the pumping rate issuperimposed to it. Recharge is about 80 mm/year. A large portion of the recharge flows through the altered granite and discharges into streams. Only a small portion of the recharge infiltrates into the granite rock mass. Regional flow within the rock mass is almost vertical. Local flow is affected by discharge into the mine gallery and the La Tarica stream, and by the presence of high permeability-preferential flows: fractures F1 and F2 and the uranium quartz vein. The hydraulic conductivity of rock mass is considered to be anisotropic.

The models used for the interpretation simulate the altered granite, the uranium quartz vein and the fractures F1 and F2 by means of two-dimensional finite elements embedded within an anisotropic three-dimensional domain (fully 3D elements) that represents the granite rock mass. This approach causes the tracers to travel with higher velocity through the major fractures (F1 and F2) and with lower velocity through the minor fractures of the granite rock mass. For some models a double porosity approach was used to simulate the diffusion of tracer into the cristalline rock. The double porosity approach assumes the medium to consist of a porosity with immobile water, where tracers are only transported by diffusion, and a porosity with mobile water, where they are also transported by advection and dispersion. The modeled domain is located between the uranium quartz vein in the South and the limit of the watershed in the North, near borehole S14. The eastern, western and lower boundaries were chosen in such a way that the influence of pumping in borehole S2 is negligible. The top layer coincides with the phreatic level.

Numerical Interpretation

We interpreted numerically breakthrough curves of the tracers recovered at borehole S2: gadolinium, uranine and deuterium, with flushed injections, and phloxine and rhenium, without flushing. The tests were interpreted again using TRANSIN-III in a three-dimensional domain. The finite element grid was made by building prisms between layers of two-dimensional triangular grids, as in the previous case.

The interpretation was performed in two steps: First, flow parameters were estimated by a steady state flow model (with pumping in borehole S2), from pressure heads measured after three weeks of pumping. In this way, a velocity field was obtained to be used afterwards for the estimation of transport parameters from measured concentration of the breakthrough curves in borehole S2. So, flow parameters and transport parameters are estimated separately. While this choice is less effective than joint estimation, we were forced to do it separately because of CPU time constraints. Table 5 show the parameters returned by the flow model.

Parameter	
K in N-S direction (m/d)	0.16e-5
K in vertical direction (m/d)	0.86e-4
K in E-W direction (m/d)	0.15e-4
T of the phreatic surface (m^2/d)	0.26e1
T of UQ vein (m^2/d)	0.12
T of fracture F1 (m^2/d)	0.48e-2
T of fracture F2 (m^2/d)	0.90e-3
Recharge to the phreatic surface (mm)	222.3
Constant level at La Tarica stream (m)	1.01
Constant level at bottom (m)	piezometry
Pumping rate (m^3/d)	-0.112

Table 5. Flow parameters returned by the numerical model of test at 2-13-15.

Following we present the full modeling process. It is not divided into first and second injection given that some feedback was produced between both models. As a consequence, we ended up with different conceptual models that explain with certain success the observed breakthrough curves.

The models for all the tracers were calibrated by estimating the longitudinal dispersivity α_L of the granite rock mass and of the fractures, the product of porosity times retardation (ϕR) of the granite rock mass, the product of the porosity times aperture times retardation ($\phi b R$) of the fracture and, in case of double porosity, the immobile porosity (ϕ_{im}) and the diffusion coefficient divided by the square thickness of the immobile porosity (D_{im}/b_{im}^2). As the transversal dispersivities α_T had little influence, they were kept at a fixed value.

Measured concentrations in borehole S13 were used to calculate the mass of tracer leaving the borehole as a function of time. For uranine and gadolinium this led to significant differences between the total amount of tracer, calculated from the measured concentration, and the amount actually used. Possible explanations for this difference include: loss of tracer from the borehole before the start of the flushing, adsorption of the tracer to the equipment walls and/or an insufficient mixing within the borehole, leading to measured concentrations that are not representative. The sudden drop in rhenium and phloxine after 30 days is due to a change in flowrate. Since the flow model is steady state, concentrations after this change are not taken into account.

In order to obtain good fits (Figure 4) for the tracer tests without flushing (phloxine and rhenium) by means of models with single porosity, we were forced to estimate also the injected mass. The difference in estimated and actual injected mass can be explained by errors in the velocity field and/or by diffusion into the cristaline rock. Indeed, acceptable fits could be obtained without estimating the injected mass. However, by means of models with double porosity, these fits are not as good as the fits of the single porosity model. Nevertheless, we feel that the double porosity model is the best one, because it reflects in a better way the processes that we think are occurring in reality, which leads to parameter values with more physical meaning and to higher predictive capabilities of the models.

Tracers injected by flushing are modeled by means of two steady state flow models: one for the period of flushing (approximately one day) and one for the period after flushing. The problem of this approach is that we could not use the algorithm of automatic calibration, because the parameters are distributed over two models. Therefore, we first calculated the breakthroughs with the transport parameters obtained from the test of phloxine, both for the single porosity and double porosity model. As it can be seen from Figure 5, the results of the double porosity model resemble better the observed data than those of the single porosity model. Then, a manual calibration (i.e., trial and error) was carried out. Taking into account that a manual calibration generally does not give as good fits as automatic calibration, rather acceptable fits could be obtained (Figure 6) by changing slightly some parameters (Table 6). It is expected that we can obtain better fits by means of an automatic calibration. For the test of uranine we multiplied the porosities of both the rock mass and the fractures by 1.5 and for the test of gadolinium by 0.6. Note, that these differences are much smaller then those obtained by the models that do not simulate the flushing of water. For the test of deuterium we had to reduce the injected mass. A possible explanation for this reduction is diffusion into the cristaline rock, which may be more important for this tracer than for the others.

Conclusions

In summary, good fits could be obtained by means of relatively simple models. However, these gave estimated parameter values with little physical meaning and have therefore little predictive capabilities. More complicated models can overcome these problems, but may encounter difficulties with respect to software and CPU time and memory. In any case, the interpretations have undoubtedly shown that models can be used to detect what kinds of processes are important for the behaviour of the tracers in groundwater. For the tests performed in boreholes S2, S13 and S15, diffusion into the cristaline rock is an important process and the influence on the groundwater flow of the flushing of water cannot be neglected (in case of the test with flushing). As seen in the S11-S12 test interpretation, tailing effects early attributed to physical-chemical processes, may appear explained by the heterogeneity. When accounting properly for different hydraulic properties of both fracture and rock mass-instead of fracture solely-, part of the tails are reproduced thanks to the late arrivals of mass travelling through less pervious medium. In fact, if only heterogeneity would be the cause of tailings, they would appear exactly the same for all tracers, which is not the case. The apparent dependence of the tail and peak retardation with respect to molecule size, seems to point out that matrix diffusion is more important than adsorption. Besides, the injection procedure also affects the type of breakthrough, and this effect, plays an important role in controling mass dilution in the rock mass and therefore, the extend to which retardation processes may appear.

One of the main conclusions from the modeling exercise was that the most meaningful results were obtained with models that reproduce the 3D nature of the flow system. Simplified models led to parameters of doubtful validity for prediction purposes. Predictive capabilities of the 3D models were verified by simulating tests under flow conditions different from those used for calibration. It was shown that both single and double porosity models led to fair predictions, but that the latter was superior. Finally, it was also shown that reasonable parameters and model fits could only be obtained when processes often neglected in normal permeability media (natural flow field, flushing, etc) were included in the interpretation.

Parameter	Injections without flushing		Injections with flushing		
	Phloxine	Rhenium	Gadolinium	Uranine	Deuterium
Single porosity model without flushing					
α_L rock (m)	0.16e1	0.13e1	0.15e2	0.13e2	0.21e2
α_L fract (m)	0.27e1	0.26e1	0.15e2	0.15e2	0.23e2
ϕR rock (-)	0.88e-3	0.72e-3	0.12e-3	0.15e-4	0.33e-3
ϕbR fract (m)	0.14e-3	0.13e-3	0.10e-3	0.13e-2	0.75e-3
Mass	63%	65%	-	-	-
Single porosity model with flushing					
α_L rock (m)			0.16e1	0.16e1	0.16e1
α_L fract (m)			0.27e1	0.27e1	0.27e1
ϕR rock (-)			0.53e-3	0.13e-2	0.88e-3
ϕbR fract (m)			0.81e-4	0.20e-3	0.14e-3
Mass				54%	
Double porosity model without flushing					
α_L rock (m)	0.10e1	0.10e1			
α_L fract (m)	0.10e1	0.10e1			
D_{im}/b_{im}^2	0.40e-4	0.40e-4			
ϕR rock (-)	0.32e-4	0.27e-4			
ϕbR fract (m)	0.33e-4	0.37e-4			
ϕ_{im}	0.45e-1	0.40e-1			
Double porosity model with flushing					
α_L rock (m)				0.10e1	
α_L fract (m)				0.10e1	
D_{im}/b_{im}^2				0.14e-4	
ϕR rock (-)				0.32e-4	
ϕbR fract (m)				0.33e-4	
ϕ_{im}				0.75e-1	

α_L = Longitudinal dispersivity ϕ = porosity R = retardation b = fracture aperture ϕ_{im} = immobile porosity b_{im} = thickness of the immobile porosity; Mass: reducing factor applied to input mass to achieve good fits. Double porosity model with flushing was only applied to uranine

Table 6. Transport parameters returned by the numerical model of test at 2-13-15.

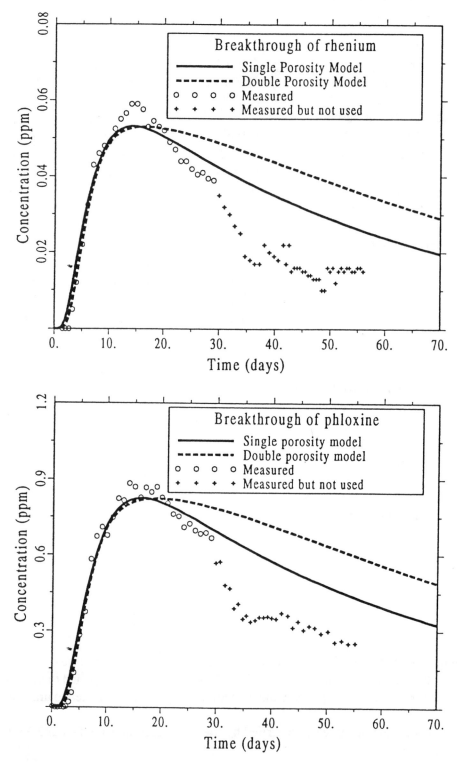

Figure 4. Calibrated breakthrough curves of rhenium and phloxine by means of 3D heterogeneous model.

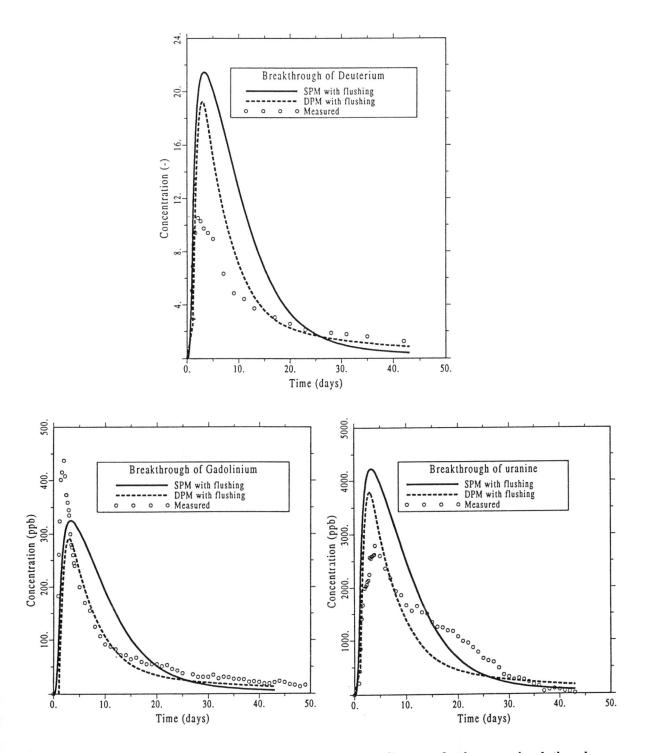

Figure 5. Breakthrough curves of flushed injected tracers. Computed values are simulations by using parameters calibrated from rhenium test. Flushing period is accounted by coupling two steady-state flow models.

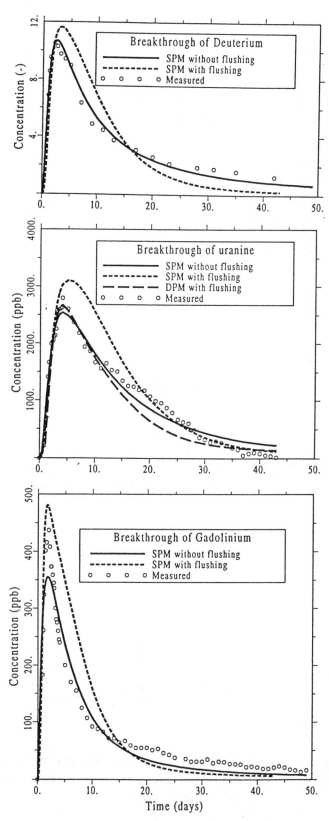

Figure 6. Separated calibrations of the tracers injected by flushing. Note that although fits are superior than those at figure 5, returned parameters are less consistent. Parameters are different for all them, since chemical behavior of the tracers is not considerered.

Acknowledgements.

Tracer tests and the whole project of El Berrocal, were funded by ENRESA and the CEC. Authors thanks all those individuals that made possible the design, performance and interpretation of these tests. Special thanks are devoted to Marco d'Alessandro (JRC, Italy) Pedro Rivas and Benigno Ruiz (CIEMAT, Spain) and José Bueno (CEDEX, Spain).

References

[1] Rivas, P. and 17 more contributors (1996) El Berrocal project. Final summary report. European Comission, Nuclear Science and Technology (in press)

[2] d'Alessandro,M.; Mousty, F.; Bidoglio, G.; Guimerà, J.; Benet, I.; Sánchez, X.; García, M. and Yllera, A. (1996) Field tracer experiments in low permeability-fractured medium: results from El Berrocal site. J. Cont. Hydrol. (in press)

[3] García, M.; Guimerà, J.; Yllera, A.; Hernández, A.; Humm, J. and Saaltink, M. (1996) Tracer tests at El Berrocal site. J. Cont. Hydrol. (in press)

[4] Benet I. and Carrera J. (1992) Desarrollo de un programa de ordenador para la interpretación de ensayos de trazadores. Hidrogeología y Recursos Hidráulicos, XVII, pp. 335-349.

[5] Sánchez-Vila, X. and Carrera, J. (1996) A solute transport equation in heterogeneous media without a fickian macrodispersive term. Submitted to Water Resour. Res..

[6] Galarza G.; Medina A. and Carrera J. (1996) TRANSIN-III. User's guide. El Berrocal project, Topical Report 16, ENRESA.

[7] Carrera J. et al. (1996) Hydrogeology task group final report. El Berrocal project, ENRESA, Madrid.

[8] Guimerà, J.; Vives, L.; Tume, P.; Saaltink, M and Carrera, J. (1996) Numerical modeling of pumping tests in low permeability and fractured medium. El Berrocal project, Topical Report 14, ENRESA

SESSION IV

Aims and Design of Planned Field Tracer Experiments
Chairmen: W. R. Alexander (University of Berne, Switzerland)
and G. Volckaert (SCK-CEN, Belgium)

A Tracer Experiment at the Kamaishi Mine As a Part of An Integrated Approach to Geosphere Transport Modeling -

Masahiro Uchida, Hiroyuki Umeki and Hidekazu Yoshida

Power Reactor and Nuclear Fuel Development Corporation (PNC), Japan

Abstract

This paper presents an overview of PNC's integrated approach to geosphere transport modeling. The approach is a complementary program of field tracer transport experiments (FTTEs), laboratory experiments, geological studies, natural analogue and, performance assessment calculations. The FTTE is being performed at the Kamaishi mine, which is located in the northern part of Honshu, the Japanese main island. As a part of the performance assessment (PA) to be completed about the year 2000, PNC is developing a geosphere transport model at the block scale. The emphasis on block scale transport is a result of PNC's first performance assessment (H3 Report), which indicated that a small region of the host rock could provide acceptable barrier performance. PNC's goal for the next PA is then to implement a more realistic geosphere transport model to enhance the confidence in the result of H3 Report. For this purpose, PNC is developing and testing a discrete fracture network model. PNC considers that two efforts need to be undertaken to build confidence in this approach. One is to develop a realistic structural model, such as the one being developed for the Kamaishi mine. The other is to develop an understanding of the flow and transport processes which are relevant to radionuclide migration.

PNC is currently preparing a borehole array to conduct non-sorbing tracer experiments. Five of the seven planned boreholes have been drilled. Spacings of holes range from 2 to 25 meters to allow tracer experiments at various scales. Two meter spacings were set for tracer experiments through single fractures. Hydraulic characterization so far has identified six highly isolated hydraulic domains (compartments). The isolation is shown by different pressures measured using multi-packer piezometers. Each domain is internally well-connected, but poorly connected to adjacent domains. This compartmentalization could potentially be a favorable hydraulic feature for PA, if found at the eventual site. Single conductive fractures will be identified and specially designed piezometers will be placed by March 1997. Tracer experiments will be completed by March 1998.

FTTEs have inherent limitations. For example, the results may be non-unique as several processes may produce similar breakthrough curves. Furthermore, FTTEs are often too short in duration to identify slow, safety relevant processes, such as matrix diffusion. A basic concept here is to employ laboratory experiments and natural analogue studies to complement FTTEs.

PNC is planning to conduct laboratory experiments at various scales, which can accommodate critical structures (heterogeneity) such as channels, fracture filling materials, and fracture intersections. Channels may accommodate several diffusive processes such as diffusion into the matrix, diffusion into stagnant pore space, and diffusion into fracture fillings. Diffusion into fracture filling could be of importance in a PA context, since more rock matrix could be accessible through fracture fillings, as indicated by alteration haloes along fracture surfaces.

Fracture intersections could be important because they may form a bypass to channels. Natural analogue studies could provide insight into diffusion over appropriate scales of length and time.

1. Introduction

1.1 The H3 Performance Assessment

In 1993, PNC completed its first performance assessment called H3[1]. The H3 report concluded that a strong engineered barrier system combined with compatible host rock could effectively retard radionuclide migration. This retardation did not require a large rock volume of the host rock.

The H3 geosphere transport model represented the fractured media in the vicinity of the disposal tunnel using equally-spaced parallel fractures. The spacing of parallel fractures was set to ten centimeters based on geological observations. The fracture surfaces were uniformly open to flow. The modeled region was varied from 10 meters to 1000 meters to look at the effect of migration distance. An equivalent hydraulic conductivity of 10^{-8}cm/s was assumed. A hydraulic gradient of 0.05 was derived from a regional flow model and applied to the geosphere transport model. The model simulated flow and transport using advection, dispersion, and matrix diffusion.

The calculation results showed that even ten meters of host rock could provide significant retardation. A small volume of undisturbed host rock in the direct vicinity of the repository may behave as an effective barrier and more distant rock volumes may serve as redundant barriers[2]. If this is the case, then detailed characterization may be required only for the undisturbed host rock in the direct vicinity of the repository. The larger rock volume outside may require less detailed characterization. This could significantly reduce the effort for site characterization.

1.2 The Approach for the Second PA

One of the critical requirements to enhance confidence in the result of the H3 performance assessment is to apply a more realistic geosphere transport model reflecting the heterogeneous features and properties of the geologic media. The second PA will be focused on a relatively small scale, such as a few tens of meters from the disposal tunnel. In order to achieve this goal, PNC is planning the following activities:

- To develop and apply a discrete fracture network (DFN) model to more realistically model heterogeneity,

- To strive for linking the detailed research model (DFN model) to the geosphere PA models in order to achieve both more realistic PA and computational efficiency,

- To conduct FTTEs at various scales ranging from a single fracture up to fracture networks of several tens of meters in size (block scale), to provide a basis for a conceptual representation of the geosphere transport model, and

- To develop an integrated approach utilizing laboratory experiments, natural analogue studies, and geological characterization to complement FTTEs. Laboratory experiments at various scales, which accommodate process specific heterogeneity, will be conducted. These laboratory experiments will focus on processes operating within single fractures and small fracture networks under well-known, controlled boundary conditions.

In summary, the Kamaishi experiments are part of an integrated program of experimental activities to understand flow and transport in fractured rocks. The FTTEs emphasize the understanding of in-situ block scale hydrogeological structure.

1.3 Rationale and objectives of Field Tracer Transport Experiment at the Kamaishi Mine

PNC is planning to conduct a series of tracer experiments at the Kamaishi mine for the following reasons. To:

- obtain a conceptual model with realistic geometries and properties. Flow and transport parameters will be determined for well-defined hydrogeologic features over a range of scales including single fractures and fracture networks, and

- understand the hydraulic properties and geometries of barriers to flow, and their effect on hydraulic gradient.

The result of the experiment will be used:

- as a basis for a conceptual representation of the block scale geosphere to be used in the next performance assessment report to be prepared about the year 2000, and

- to build confidence in application of the DFN model through the testing of the model against in-situ experiments.

The experiments further serve the purpose of (1) testing site characterization methods for use in the development of an eventual Japanese geologic repository, and (2) understanding the size of compartments bounded by barriers to flow which may influence the repository layout and seal design.

2. Field Tracer Transport Experiment at the Kamaishi Mine

2.1 Site Description

PNC is currently planning FTTEs in the Kamaishi mine, which is located in the northern part of Honshu (the main island of Japan) (Figure 1). The in-situ experimental area is located in Cretaceous granodiorite with an overburden of approximately 300 meters.

PNC initiated site characterization work at the tracer experiment area in 1994. Approximately 2,400 square meters of rock have been explored by a horizontal array of five boreholes whose lengths vary from 80 to 100 meters (Figure 2). The boreholes were drilled at various spacings between 2 and 25 meters to study features ranging from single fractures to fracture networks. A 2m spacing was adopted so as to intersect a single fracture or single fracture segment (part of a fracture without intersection/branch). Furthermore, the 2m

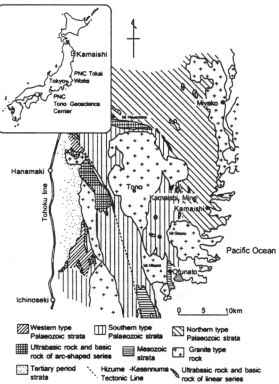

Figure 1. Location and geology around the Kamaishi mine

spacing borehole array may be used to test closely-spaced lower transmissivity fractures, while the larger scale borehole array may be used to test the less frequent, higher transmissivity fractures.

2.2 Approach to Characterizing Hydrogeologic Structures

A sufficient understanding of the hydrogeologic structure is a prerequisite for a successful tracer testing program. Hydrogeologic characterization has been implemented in the following two stages:

1) Identification of zones (compartmentalization), and
2) Identification of conductive fractures.

The experiment area shows highly isolated hydraulic domains which are indicated by different pressures measured by multi-packer piezometers. The knowledge of these compartments is important for tracer test design, and the compartmentalization is a potentially favorable feature for PA (barrier to flow). We are developing this understanding by carefully monitoring the hydraulic conditions of the site throughout drilling and testing. The approach includes the following steps:

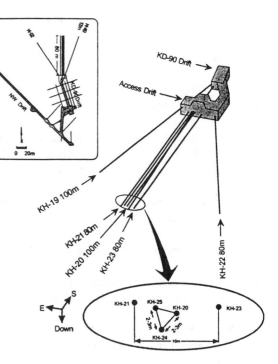

Figure 2. Borehole array of the tracer experiment area in the Kamaishi mine

1) Identification of zones (compartmentalization)
- continuously monitoring pressures in boreholes using seven to eight packer piezometer systems,

- automatically monitoring drilling depths of new holes to identify precisely the location of fractures which cause pressure interference in other boreholes,

- logging fluid flow within the boreholes at a resolution of 1 meter or better, and

- building a geologic model of connections using television and geophysical logs.

2) Identification of conductive fractures

- confirming the locations of crosshole connections using simple pressure interference and tracer tests, and

- using the drilling response, flow logs, and crosshole tests to design final multi-packer piezometer systems for tracer testing.

2.3 Hydrogeologic Structure of Test Site

The characterization work has identified six distinct hydrogeologic zones based on measured groundwater pressures, pressure responses during the drilling, and pressure build-up testing[3]. These zones approximately vary from a few hundreds to over one thousand square meters in plan area. The flow system is highly compartmentalized. Each zone is well-connected internally and poorly connected to other zones. The locations of zones and their initial pressure

194

are shown in Figure 3. Barriers to flow may be explained by either heterogeneity of hydraulic properties among fractures or by network geometries which are highly clustered.

2.4 Tracer Testing Plans

Preliminary pressure interference and tracer tests are planned for later this year. The results of the preliminary tracer tests will be used to plan the full scale tests to be run in the following year. The detailed tracer experiment design has yet to be decided. However, the concept can be described as follows:

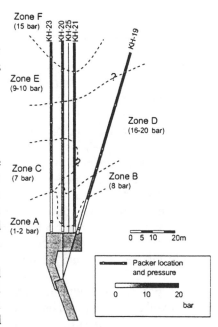

Figure 3. Pressure map of tracer experiment area

- only non-sorbing tracers (possibly saline tracers) will be used, since the experiments have to be terminated by March, 1998. An emphasis will be placed on tracers that can be analyzed at the site and monitored downhole,

- effort will be concentrated on defining transport properties within individual zones and across barriers to flow,

- tests will be run over various distances ranging from 2 to 50 meters, shorter scales being used for single fracture tests and larger scales for fracture network tests, and

- tests will be run using either convergent flow or variable strength dipole arrangements.

Each section of the multi-packer piezometers includes a flow line and a separate pressure measurement line. Critical zones for tracer testing (up to three per hole) have an additional flow line and sensors to measure temperature and fluid electrical conductivity. Saline tracer concentration will be measured downhole, as well as at the hole collar. Each piezometer has a uniform 70-mm diameter, installed within a 76-mm hole, to minimize the storage volume in the test intervals.

2.5 Conceptual Model Development

The DFN model will be used to develop the conceptual model. The basic approach identifies pathways beginning with single fractures and then progressively enlarges the scale of investigation to include fracture networks of increasing scale. The possible modeling steps are as follows:

Single Fracture:
A set of tracer tests will be conducted among the 2m spacing central array boreholes, KH-20, 24 and 25, since this array is most likely to capture single fractures or single fracture segments. At least one single fracture segment will be selected and packed off as one of the three specially instrumented zones of each borehole. A test will be run to identify the effective porosity and dispersivity. The test will be modeled with the single fracture feature within a DFN model or with a simple advection/dispersion model with matrix diffusion. Also, a single fracture test will be conducted by using larger-scale borehole arrays to study the scale dependency of dispersivity, if an appropriate fracture can be found. Process models consistent with laboratory experiments described in Section 3.1 will be used to simulate the tracer experiments.

Fracture Network:

A fracture network model will be developed encompassing the different scales of the borehole array. The fractures in the models will be conditioned to the locations, orientations, transmissivities, and other properties of the known conducting features. The properties may also be conditioned to the results of crosshole interference tests as a means of realistically reproducing the hydraulic connectivity.

The smallest region enclosed by KH-20, 24 and 25 could contain some simple fracture intersections, as well as single fractures. The structure of the fracture system will be deterministically defined to the extent possible through television logs, pressure interference tests and single hole radar measurements. Conditionally generated DFN models will be developed using varying degrees of constraints by changing data categories to be used. Predictions based on these models will be compared with the deterministically defined fracture system and the measurements of the tracer breakthrough curves. Critical data/tests for conditional modeling can be identified precisely, for this scale, based on sufficient information. The mixing rate at the fracture intersection could be studied by tracer experiments.

For a larger region, such as the rock mass between KH-22 and 23, three kinds of DFN models will be developed using different numbers of boreholes to constrain the models. The least constrained model uses only the information from the central array boreholes, KH-20, 24 and 25. This case looks at the model's capability to extrapolate conditions to larger scales. The second case uses only the information from the tracer test holes, KH-22 and 23. The third case, which is most constrained, uses the information from all boreholes, KH-20, 22, 23, 24 and 25. The tracer test between KH-22 and 23 will then be predicted by using these three models respectively. By comparing these three model predictions with the measurements of the tracer breakthrough curves, we can determine the predictive value of the testing and modeling information. The efforts can help to optimize future site characterization approaches.

Information on transmissivity and mixing rate of fracture intersections obtained from NETBLOCK (Section 3.1) will also be used to simulate the results.

2.6 Building on Past Field Tracer Transport Experiments

The bulk of the past tracer test work has focused more on single features (for example, Stripa H-zone[4], Grimsel MI experiment[5]) and less on fracture networks. The Kamaishi tests will be running at about the same time as similar network-scale experiments in the Swedish Äspö Hard Rock Laboratory (TRUE)[6] and the Canadian URL (moderately-fractured rock test)[7]. Thus, the focus on fracture networks and compartments is timely and challenging.

The Kamaishi program is not intended to develop new tracer test methods. Rather, we expect to draw on the experience developed from Stripa, Grimsel (both MI[8] and BK experiments[9]), and the Canadian URL (highly fractured rock experiments[10]).

Understanding hydrogeologic structure and boundary conditions is a prerequisite for successful tracer experiments. Site preparation work at the Kamaishi mine has been clarifying compartmentalization and conductive features within compartments as described in Section 2.3. Thus it should be relatively easy to establish a simple flow field during the tracer tests.

PNC regards the Kamaishi program as a part of an integrated study including laboratory studies, natural analogue studies, and other supporting studies.

3. Integration of Field Tracer Transport Experiment and Other Studies

PNC's focus is to build confidence in the DFN approach for modeling block scale flow and transport, since radionuclide migration at this scale is controlled by discrete fractures. Therefore the PNC research and development program aims at an integrated effort which includes laboratory testing in single and multiple fractures, flow and transport tests in domestic underground facilities, natural analogue studies, geological studies, and cooperative studies in international underground test facilities. The results of the effort support the development of numerical simulators of block scale flow and transport. Finally, development of linkage between the DFN model and geosphere PA models is undertaken. These, in turn, support performance assessment models that take into account the processes which significantly affect waste disposal safety.

3.1 Expected Processes in Each Scale and Planned Laboratory Experiments

There seems to be insurmountable difficulty in identifying transport processes when only relying on breakthrough curves measured via FTTEs. One source of difficulty, is poor understanding of hydrogeologic structure (heterogeneity). Therefore, we are exploring how to run complementary laboratory experiments whose scales are suitable to include the specific heterogeneities of interest, such as channels, and fracture intersections. We are currently dividing structure and processes operating within fractured media in the following hierarchical manner and preparing laboratory experiments at various scales which include the relevant structures (heterogeneity). An example of the relationships between structures and processes are shown in Figure 4.

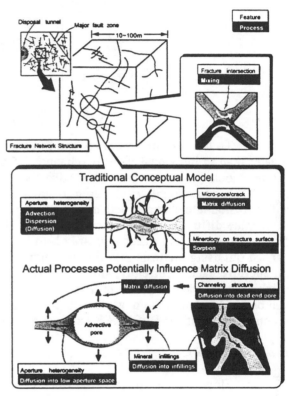

Figure 4. An example of relationships between transport processes and structure within fractured media

Single Fracture:

From geological observations, areas along individual fractures can be divided into the following two components, *open channel area* and *fracture filling materials area*. *Open channel areas* may be further divided into *advective channel areas* where advective flow dominates, and *diffusive channel areas* where the water is stagnant due to either narrow aperture or a lack of connectivity. Diffusion may be important as a transport mechanism in these stagnant areas. *Diffusion through fracture filling materials* is potentially very important for long-term PA, since this diffusion could provide additional access to the rock matrix. In other words, this diffusion could significantly increase fracture surface area accessible to rock matrix[11]. This is indicated elsewhere by development of alteration halos along whole fracture surface.

Since the width of advective channels in fractures is often less than a few tens of centimeters, scales less than this could be represented as *open and parallel plates (with heterogeneity)*. This scale is suitable to test traditional processes within so-called parallel plates, such as advection, dispersion, and matrix diffusion. In the laboratory, PNC is planning to conduct small-scale column (5-cm diameter) experiments on such features.

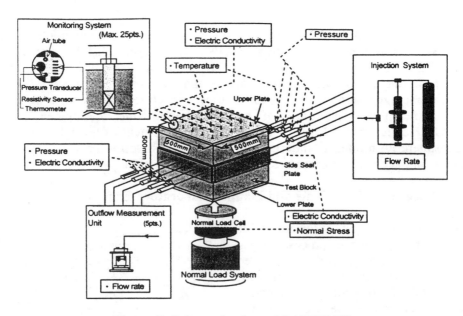

Figure 5. Schematic view of LABROCK

To test *diffusive channels* and *fracture filling materials*, the experiment scale should be greater than an individual advective channel. One set of experiments, known as LABROCK, uses a 50-cm block [12]. An individual advective channel can be encompassed at this scale. The LABROCK experiments will be used to study advection and dispersion within advective channels, and diffusion into channels with narrow aperture or stagnant pool. LABROCK may also be able to investigate diffusion into fracture filling materials through long duration tests. Currently, flow tests and non-sorbing tracer tests are planned. A detailed study on aperture distribution will be carried out using the resin injection technique. Flow wetted surface areas will be evaluated based on aperture data. Currently, an artificially made rough fracture has been tested. Experiments using a natural fracture will be initiated next year. The in-plane heterogeneity of fracture apertures will be modeled with the newly developed in-plane heterogeneity feature of the DFN model. A schematic view of the LABROCK experiment is shown in Figure 5.

Fracture Network:

Flow and mass transport in fracture networks may be more complex than flow and mass transport in a series of connected single fractures. Specifically, several papers have reported that fracture intersections have increased transmissivity[13,14,15]. Also, mixing at fracture intersections could add further complexity.

The NETBLOCK experiments will use blocks as large as 1 m³ containing multiple fractures. The first NETBLOCK experiment will use a rock block which contains intersecting double fractures in order to look at an intersection as a possible critical pathway and to study mixing behavior at fracture intersections. A block containing a fracture network will be tested after completion of the double fracture test. NETBLOCK is now at the final design stage and construction of the equipment will be completed in March, 1997. A schematic view of the NETBLOCK experiment is shown in Figure 6.

3.2 Natural Analogue

Natural analogue study might be useful to understand slow processes such as matrix diffusion. Natural uranium-series disequillibrium within an alteration halo and adjacent intact rock have been studied[16]. The disequillibrium suggests that Uranium has migrated from the fracture surface to the halo region over geologic time. These results could provide information on the matrix diffusion depth. Uranium concentrations were also measured and the higher uranium concentrations within fracture fillings indicates significantly higher retardation.

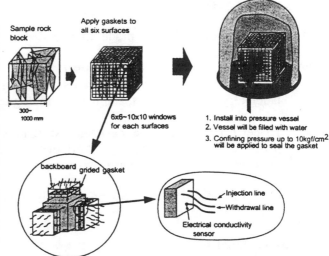

Figure 6. Schematic view of NETBLOCK

3.3 Geological Characterization of Water Bearing Fractures and Other Supporting Laboratory Measurements

Geological observation at the Kamaishi mine have identified three types of fractures (Type A, B and C, Figure 7)[17]. Fractures are classified in terms of width of fracture filling, existence of the alteration halo, and texture (crushed zone, fault gouge). This information could provide the basis for addressing representativeness of the fractures to be used for tracer tests. It will also form a basis for synthesizing the measurements and assigning properties for the block scale DFN model which is being constructed based on a set of parameters determined for reliably defined fracture groups.

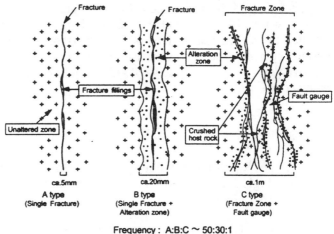

Figure 7. Type of fractures at the Kamaishi mine

Also, aperture data (channel geometry) will be obtained in-situ by excavation of fractures after being impregnated with resin. This data could be useful for addressing the representativeness of LABROCK samples in terms of dimension of channel, aperture, and fracture filling materials.

Complementary laboratory measurements have been carried out to obtain parameters necessary for modeling tracer tests. These parameters includes matrix porosity, distribution coefficient and diffusion coefficients of fracture filling material, alteration halo, and intact rock, respectively. These data have been obtained for each type of fracture.

4. Summary and Conclusions

PNC's approach to develop geosphere transport models can be summarized as follows:

- To focus on block scale geosphere transport,

199

- To apply a discrete fracture network (DFN) model to realistically represent heterogeneity at block scale,

- To strive for establishing model linkages between DFN models and geosphere PA models,

- To conduct FTTEs at the Kamaishi mine as a part of the integrated study including laboratory experiments of various scales, geological observations and supporting laboratory measurements, and natural analogue study,

- To perform laboratory experiments at various scales to study flow and mass transport processes, since controlling laboratory experiments with respect to geometry and boundary conditions is relatively easier than FTTEs, and

- To use FTTEs to derive flow porosity, dispersivity and connectivity information, and to test transport models with more emphasis on hydrogeologic structure.

5. Acknowledgment

In-situ experiment work at the Kamaishi mine is conducted by Messrs. Atsushi Sawada and Takeshi Senba of PNC, Messrs. Michito Shimo and Hajime Yamamoto of the Taisei Corporation, and Messrs. Hiroyuki Takahara and Takeyuki Negi of the Nittetsu Mining Co. Authors would like to gratefully acknowledge their dedicated and careful work. Authors are indebted to Dr. Thomas W. Doe of Golder Associates Inc. for providing important suggestions for in-situ experiment at the Kamaishi mine, and also reviewing the manuscripts and suggesting improvements. Authors also would like to acknowledge Dr. Erik K. Webb of PNC for reviewing the manuscript.

6. References

[1]   Power Reactor and Nuclear Fuel Development Corporation, Research and development on geological disposal of high-level radioactive waste, First Progress Report., 1992, PNC TN1410 93-059.

[2]   Umeki, H., H., Ishiguro, K., Hatanaka, K. and Naito, M., Near-Field Environment as an Effective Barrier against Radionuclide Transport, 1993, Proc. 1993 Int'l Conf. on Nuclear Waste Management and Environmental Remediation, Prague Czech Republic, Vol.2, p.747-754.

[3]   Sawada, A., Uchida, M., Senba, T., Shimo, M. and Doe, T.W., A Characterization of Conductive Features Using Pressure Interference Observation during Borehole Drilling, 1996, The 27th Symposium of Rock Mechanics, Tokyo Japan, p.186-190 (in Japanese with English abstract).

[4]   Birgersson L, Widén H, Ågren T, Neretnieks I, Moreno L., Site Characterization and Validation- Tracer Migration Experiment in the Validation Drift, Report 2, Part 1: Performed Experiments, Results and Evaluation, 1992, SKB Technical Report 92-25.

[5]   Frick, U. et al., Grimsel Test Site, The radionuclide migration experiment - overview of investigations 1985-1990, 1992, Nagra Technical Report 91-04.

[6]   Winberg, A., Tracer Retention Understanding Experiments (TRUE). Test Plan for the First TRUE stage, 1994, SKB HRL Progress Report, PR25-94-35.

[7]   Frost, L.H., Davison, C.C., Vandergraaf, T.T., Scheier, N.W. and Kozak, E.T., Field Tracer Experiments at the Site of Canada's Underground Research Laboratory, 1996, Proc. OECD/NEA The First Geotrap Workshop.

[8]   Frick, U., Alexander, W.R., Baeyens, B., Bossart, P., Bradbury, M.H., Bühler, C., Eikenberg, J., Fierz, T., Heer, W., Hoehn, E., McKinley, I.G. and Smith, P., The Radionuclide Migration Experiment - Overview of Investigations 1985-1990, 1992, Nagra Technical Report 91-04.

[9]   Pahl, A., Liedtke, L., Naujoks, A. and Bräuer, V., Fracture System Flow Test, 1992, Nagra Technical Report 91-01E.

[10] Davison, C.C., Frost, L.H., Kozak, E.T., Scheier, N.W., and Martin, C.D., Investigations of the Groundwater Transport Characteristics of a Large Fracture Zone in a Granite Batholith at Different Size Scales, Scale Effects in Rock Masses 93, Pinto da Cunha (ed.), 1993, Balkema, Rotterdam, ISBN, p.331-338.

[11] Olsson, O., Netretnieks, I. and Chetkovic, V., Deliberations on Radionuclide Transport and Rationale for Tracer Transport Experiments to be Performed at Äspö-A Selection of Papers, 1995, SKB Progress Report 25-95-01.

[12] Uchida, M., Noda, K., Maruyama, M. and Sudo, K., Laboratory Experiment on Flow and Mass Transport Properties of Single Fracture Rock Block, 1995, The 26th Symposium of Rock Mechanics, Tokyo Japan, p.156-160 (in Japanese with English abstract).

[13] Abelin, H., Birgersson, L., Widén, H., Ägren, T., Moreno, L. and Neretnieks, I., Channeling Experiment, 1990, SKB Technical Report 90-13.

[14] Shimo, M. and Iihoshi, S., Experimental and Numerical Study on Fluid and Mass Transport through Fractured Rock, 1995, The 8th ISRM Congress, Makuhari Japan, p.803-806.

[15] Hakami, E. and Stephansson, O., Experimental Technique for Aperture Studies of Intersecting Joints, 1993, Eurock '93, Lisbon, p.301-308.

[16] Osawa, H., Sasamoto, H., Nohara, T., Ota, K. and Yoshida, H., Development of a Conceptual Model of Nuclide Migration in Crystalline Rock, 1994, Scientific Basis for Nuclear Waste Management XVIII, Kyoto Japan, Vol.353, p.1267-1273.

[17] Yoshida, H., Amano, K., Sasamoto, H. and Senba, T., Fracture Geometry Analysis in Relation to Mass Transport in Crystalline Rock - A Case Study of Kurihashi Granodiorite at 730m Depth in Kamaishi Mine, Northeast Japan -, (in print), Journal of Geological Society of Japan (in Japanese with English abstract).

Designing a Large Scale Combined Pumping and Tracer Test in a Fracture Zone at Palmottu, Finland

Erik Gustafsson
Rune Nordqvist
GEOSIGMA, Sweden

Juhani Korkealaakso
VTT/CI, Finland

Germán Galarza
UPC, Spain

ABSTRACT

The Palmottu Natural Analogue Project in Finland was initiated in 1988, and continued as an EC-supported international analogue project in 1996, in order to study radionuclide migration in a natural uranium-rich environment. The site is located in an area of crystalline bedrock, characterized by granites and metamorphic rocks. The uranium deposit extends from the surface to a depth of more than 300 m, and have a thickness of up to 15 m.

An overall aim of the project is to increase knowledge of factors affecting mobilization and retardation of uranium in crystalline bedrock. One of the important tasks within the project is to characterize the major flow paths for the groundwater, i.e. important hydraulic features, around the orebody. This paper describes a planned experiment in one such feature, a sub-horizontal fracture zone which cross-cuts the uranium mineralization. This structure is interpreted to consist of closely located parallel fracture planes, and considered an important flow route. The zone is delineated by intersecting vertical fracture zones, where the intersections form particularly transmissive flow "channels". The objectives of the planned combined pumping and tracer test described here is to verify and further up-date the present hydro-structural model around the central part of the mineralization, increase the current understanding about the hydraulic and solute transport properties of the sub-horizontal fracture zone, as well as to verify and further characterize its hydraulic boundaries.

The detailed design is still under discussion and may be changed from the description in this paper before the final version is launched. However, the general experimental design philosophy is to use a combination of robust and well tested equipment and experimental methods/procedures, and to make the design as simple as possible, without compromising the possibilities to meet the overall objectives of the experiment. Further it should be emphasized that quantitative and stepwise modelling will be integrated throughout the sequence of tests, in order to optimize the experimental performance.

Prior to the combined pumping and tracer test, a relatively large number (10-15) of short-term cross-hole tests will be carried out, both in the subhorizontal zone as well as in adjacent vertical zones. The results are expected to validate and/or modify the current conceptual understanding of the hydraulic system. In particular, the hydraulic connections between the subhorizontal zone and some of the vertical zones will be studied. The evaluation of these tests will also form a basis for the final design and planning of the combined pumping and tracer test.

The basic experimental concept is pumping in a central borehole, while tracers are injected in a number of packed-off borehole sections that also serve as pressure measurement sections. Pressures are also monitored in all other available borehole sections during the test. Tracers will be injected when the hydraulic pressure monitoring shows that steady-state conditions have been reached. The tracers used will be non-sorbing fluorescent dyes and stable metal complexes. Tracer mass release to the hydraulically active fractures will be controlled and measured by applying a decaying pulse injection. Groundwater flow rates through the injection sections will be determined by means of dilution measurments both prior to and under pumped conditions. Increased flow rate during pumping verify flow routes connecting injection and pumping boreholes.

The test data will be evaluated in four steps: 1) exploratory analysis of hydraulic responses (drawdown curves) by response matrix and one dimensional type curve analysis, 2) hydraulic parameter estimation with a large-scale three-dimensional model (based on previous design calculation results), 3) one-dimensional analysis of tracer breakthrough curves and 4) if possible, simultaneous/combined estimation of hydraulic and transport parameters with a two-dimensional model using tracer breakthrough data and drawdown data simultaneously. Combining pumping and tracer test, with simultaneous interpretation of drawdown and tracer breakthrough curves, is a fairly new approach, and will give more information about the flow and transport properties than a hydraulic or tracer test alone.

The results of the evaluation of the combined pumping and tracer test will reveal properties of the natural flow system at the site. Such information is vital for a quantitative analysis of the forthcoming analoge transport study, which assesses the present mobilization and fracture controlled transport of radionuclides.

The interpretation of pumping and/or tracer tests is always more or less ambiguous, especially in two- or three-dimensional flow systems. In our experience, much of the ambiguity is caused by uncertainties about hydraulic boundaries to the studied system. In this project, more emphasis than before has been placed on the characterization of hydraulic boundaries to the fracture zone to be used for the tracer test.

INTRODUCTION TO THE PALMOTTU NATURAL ANALOGE PROJECT

The Palmottu Natural Analogue Project in Finland was initiated in 1988, in order to study radionuclide migration in a natural uranium-rich environment [1]. In December 1995 the Commision of the European Communities and the Geological Survey of Finland (GTK), Swedish Nuclear Fuel and Waste Mangement Co (SKB), Empresa Nacional de Residuos Radioactivos S.A. (ENRESA) and the Bureau de Recherches Géologiques et Minères (BRGM) agreed to continue this study as an multipurpose project. On January 1 st 1996 the project "Transport of radionuclides in a natural flow system at Palmottu" formally commenced, under the Nuclear Fission Safety research and training programme.

The Palmottu site is located in an area of crystalline bedrock, characterized by granites and metamorphic rocks. The uranium deposit extends from the surface to a depth of more than 300 m, and have a thickness of up to 15 m. An overall aim of the project is to increase knowledge of factors affecting mobilization and retardation of uranium in crystalline bedrock rather similar to that occuring in the proposed sites for the location of repositories for high level radioactive waste in both Finland and Sweden. Six specific objectives have been addressed for the project [2]

1) Provide a quantitative description of a Uranium-Thorium ore deposit located in a granitic host rock

2) Assess the relative importance of processes that control groundwater flow in fractured crystalline rocks

3) Study and model the effects of geochemically controlled oxidation and reduction processes on the movement of radionuclides in fractured crystalline rocks

4) Study and model the role of various retardation mechanisms on radionuclide transport

5) Study the effect of repeated glacial cycles on hydrogeological systems

6) Use data and understanding gained to develop and refine models for use in assessing the performance of potential sites and design for high-level waste repositories in crystalline rocks

RATIONALE AND OBJECTIVES OF THE PLANNED COMBINED PUMPING AND TRACER TEST

Natural analogue studies have been recognised as a very useful approach in testing models used in repository performance assessment. However, the full potential of analogue studies is not being utilized in performance assessment, since more precise and quantitative data first need to be derived from the natural analogue studies. Therefore, there is a need to better constrain the boundary conditions of processes and events relevant to performance assessment. This requires stricter physico-chemical constraints in natural analogue studies. This may be accomplished by:

• Laboratory experiments
• *In situ* experiments
• Greater use of modelling in guiding the direction of natural analogue studies

One aim of the Palmottu Project is the assessment of present mobilization and fracture-controlled transport of radionuclides. The present structural and hydrogeological model shows a sub-horizontal fracture zone, cross-cutting the U-Th mineralization, to be of major importance to the flow system around the mineralization. For the forthcoming analogue transport study the flowpaths around the orebody have to be identified in more detail. The Palmottu Natural Analogue Project has therefore been planned to integrate *in situ* tracer tests, combined with a pumping test. Combining pumping and tracer test, with simultaneous interpretation of drawdown and tracer breakthrough curves, is a fairly new approach, and will give more information about the flow and transport properties than a hydraulic or tracer test alone. The objectives of the combined pumping and tracer test are:

• Verification of updated conceptual hydrogeological model around central part of mineralization
• Identify the main potential flowpaths at the scale of the test
• Increase the understanding about flow and solute transport properties at the site

OUTLINE OF EXPERIMENTAL DESIGN

Background information relevant to the planned test

The site is located in an area of crystalline bedrock, characterized by granites and metamorphic rocks, Figure 1. The sub-vertical uranium mineralization selected for the Palmottu natural analogue study is located between the main western and eastern granite bodies and extends from the surface to a depth of more than 300 m, and have a thickness of up to 15 m [3]. The uranium mineralization is cross-cut by a sub-horizontal fracture zone, H1. This structure is interpreted to consist of closely located parallel fracture planes, and considered an important flow route. The zone is delineated by intersecting, almost parallel vertical fracture zones, V1 - V9, where the intersections form particularly transmissive flow "channels", Figure 2. The structures of the western and eastern granites together with V1 and H1 are interpreted to be the most important hydraulic conductors around the mineralized uranium rich pegmatites.

A large amount of 46 mm diameter cored boreholes provide relatively well defined structural and lithological features at the site. Within an area of approximately 300 x 300 metres 48 boreholes are arranged in ten profiles, 25 to 50 m apart. The boreholes are inclined, 20 -360 m in length reaching vertical depths of 15 to 280 m. Figure 3 shows five drilling profiles and the structures V1, V3, V6 and H1 in a 3-dimensional view. Zone H1 is penetrated by 31 boreholes. Of these have 13 been flow logged, of which 6 showed high inflow rates at the intersection with H1 zone (4 -10 l/min by airlift pumping). The hydraulic transmissivity of H1 has been calculated to range from $1.1 \cdot 10^{-6}$ to $2.2 \cdot 10^{-5}$ m^2/s.

The surface elevation at the site is within 107 - 128 m.a.s.l., Figure 4. Recharge of fresh dilute waters most likely takes place along the sub-vertical structures associated with the western granite body. The groundater flow direction of the dilute waters in the bedrock is interpreted to be towards the SSE. The driving force is the steep topographical gradient (about 50 m/km) at the southeastern border of the Palmottu bedrock block, about 1 km southeast of the studied mineralization.

Test design

The test is planned to be conducted during early summer of 1997. The detailed design is still under discussion and may be changed from the following description before the final version is launched.
The general experimental design philosophy is to use a combination of robust and well tested equipment and experimental methods/procedures, and to make the design as simple as possible, without compromising the possibilities to meet the overall objectives of the experiment. Further it should be emphasized that quantitative and stepwise modelling will be integrated throughout the sequence of tests, in order to optimize the experimental performance.

The basic experimental concept is pumping the sub-horizontal zone H1 between double-packers in a centrally located borehole, creating a converging flow field, while tracers are injected in a number of surrounding packed-off borehole sections that also serve as pressure (drawdown) measurement sections. Pressures are also monitored in all other available borehole sections during the test. The pumping phase is assumed to be approximately two weeks followed by a similar recovery period. Distances between injection borehole section and pumped borehole are foreseen to be within 100 metres. The tracers used will be non-sorbing fluorescent dyes and stable metal complexes. The tracers considered will be tested in laboratory on geologic samples from the site with synthetic groundwaters mimicing the actual water chemistry at the Palmottu site.

Prior to the combined pumping and tracer test, a relatively large number (10-15) of short-term cross-hole tests will be carried out, focused on V1, H1, western and eastern granites, which are considered most important for the analogue studies. The results from the short-term tests are expected to improve the current conceptual understanding of the hydraulic system, including the properties of the hydraulic boundaries as well as preliminary values of hydraulic parameters. The evaluation of these tests, by means of three-dimensional hydraulic models, will also form a basis for the final design and planning of the combined pumping and tracer test. Design calculations using 3-D and 2-D models will assist in choosing pumping rates, duration of test and compare alternative locations for possible tracer injection. The design calculations should also show the importance of remaining uncertainties about the system. Remanining uncertainty about the hydraulic boundaries will be especially critical to the evaluation of the long-term test and it is important to design a test that will give further information about such factors as well as about hydraulic and transport properties within the zones. The design calculations based on parameter sensitivities will also show, for a given measurement strategy, how many and which parameters that may be reliably estimated from the forthcoming large-scale pumping and tracer test.

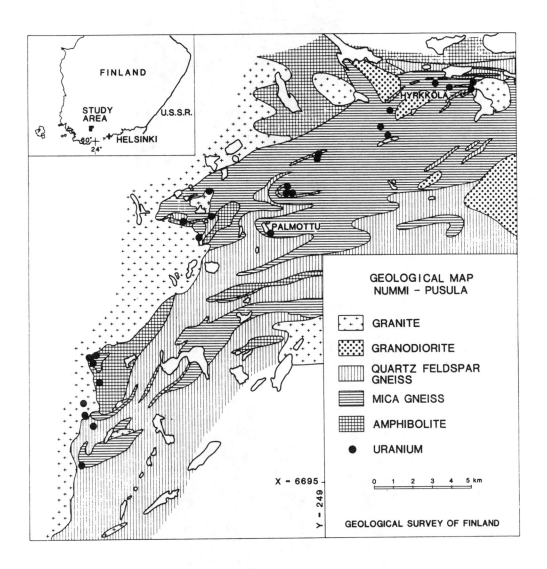

Figure 1. Geological map and the location of the Palmottu study site

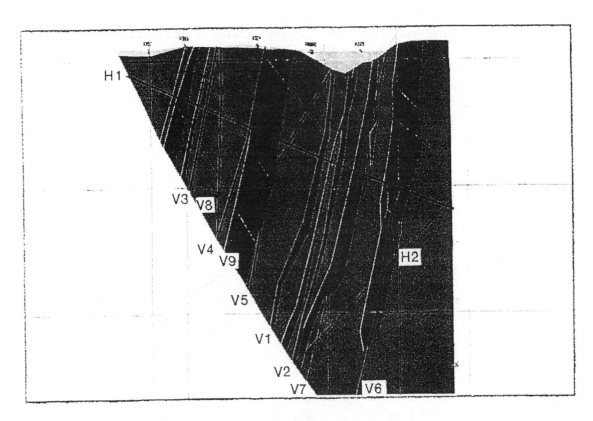

Figure 2. Two-dimensional vertical section of the structures V1 -V9 and H1 -H2

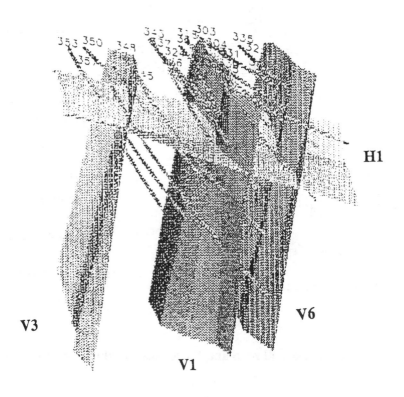

Figure 3. Three-dimensional view of the structures V1, V3, V6 and H1 with five drilling profiles

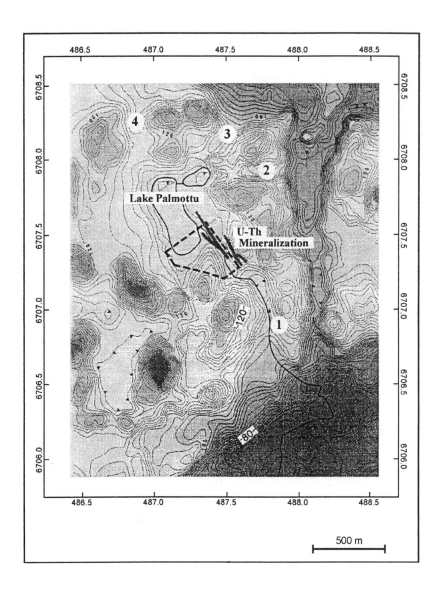

Figure 4. Topographical map, with mineralizations indicated by bold lines

Field activities will start by drilling the centrally located 5" diameter pumping borehole and three to six 4" tracer injection boreholes 100 - 150 m deep, at the locations most favourable accordning to the design calculations. Water sampling for chemical characterization and limited hydraulic testing will be performed during drilling in order to get first strike samples related to specific hydraulic features. The new boreholes will thereafter be hydraulically tested in more detail by means of spinner measurments or water injection tests. Based on the test results packer positions will be decided for tracer injection and pumping. The tracer injection boreholes will be instrumented with a double-packer system for simultaneous tracer injection, groundwater flow measurement and pressure monitoring, Figure 5. The tracer injection sections will be manually sampled throughout the experiment for continous monitoring of the groundwater chemical composition. The pumping borehole will have automatic flow regulation and a pressure transducer in the pumped section. The pumping borehole will also be equipped with an automatic water sampler, valves for manual sampling, electrodes and monitoring devices for pH, Eh and electrical conductivity. The strategy for pumping and tracer sampling will depend on the hydraulic properties of the zone interval in the pumped borehole [4]. For example, it may be desirable to obtain tracer samples from more than one conductive part of the zone.

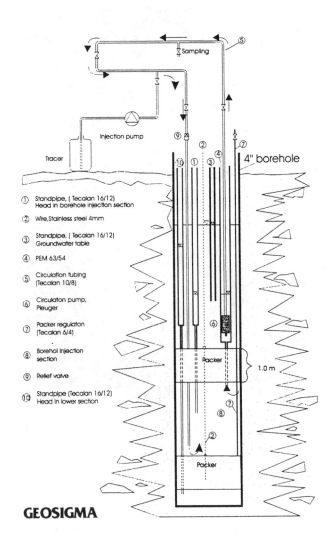

Figure 5. Proposed equipment set-up in tracer injection boreholes

To ensure a successful pumping and tracer test, hydraulic connections, especially to tracer injection boreholes, and borehole installations will be checked in advance by pumping the 5" borehole for about 24 hours and measure drawdown in the surrounding boreholes. Groundwater flow rates through the tracer injection borehole sections will be determined during this pumping by means of dilution measurments. When hydraulic gradients have reached natural conditions again after the 24-hours pumping, groundwater flow rates will be measured once again through the tracer injection sections and compared with flow rates under pumped conditions. Increased flow rate during pumping verify flow routes connecting the injection and pumping boreholes. Considering tracer detection limit and dilution in pumping borehole, a flow rate above about 10 ml/min is also required if tracer injection techniques without excess pressure is to be applied in a tracer test of the Palmpottu scale. A valuable experience from previous tests at the Hard Rock Laboratory at Äspö, Sweden [5] and El Berrocal, Spain [6] was that groundwater flow determinations, using dilution measurements, before and during pumping were very useful for identifying hydraulically connected parts of the flow system, Figure 6.

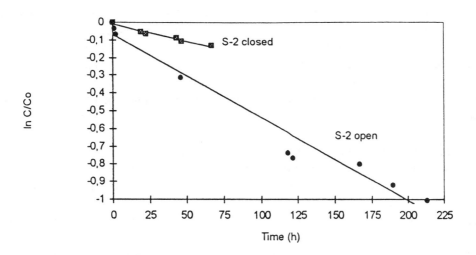

Figure 6. Point dilution test result in S-13 upper section, El Berrocal

Before the pumping will start for the combined pumping and tracer test, final design calculations will be made, incorporating the results obtained from testing of the recently drilled large diameter boreholes, and decisions will be taken on pump flow rate, tracer injection proceedures, mass and concentration of injected tracers and sampling schedules. After start of pumping, the tracers will be injected when the hydraulic pressure monitoring shows that steady-state conditions have been reached.

Even though it is common knowledge that the input function must be known for a successful interpretation of a tracer test, many tests are still performed with almost unknown source terms. In this test tracer mass release to the hydraulically active fractures will be controlled and measured by applying a decaying pulse injection where a tracer pulse is injected without excess pressure into the borehole section and then released to the fractures by the ambient groundwater flow through the section. However, if it of some particular reason is decided to inject tracer at a location where groundwater flow rate through the borehole section is very low, chase-fluid will be used to flush out the tracer from the injection section.

Although all experimentally obtained data will be handled and stored according to a quality plan, an important lesson learnt from previous tracer tests is the need for a comprehensive documentation of the test. In a "Log of Events" all events and activities within the frame of the actual experiment will be noted, as well as other activities at the site and nearby locations. Ideally, manmade mistakes and outer disturbances will then not be interpreted as properties of the studied fracture zones.

EXPECTED OUTCOME AND USE OF RESULTS

The evaluation of the test data involves the following steps: 1) exploratory analysis of hydraulic responses (drawdown curves) by response matrix and one dimensional type curve analysis, 2) hydraulic parameter estimation with a large-scale three-dimensional model (based on previous design calculation results), 3) one-dimensional analysis of tracer breakthrough curves and 4) if possible, simultaneous/combined estimation of hydraulic and transport parameters with a two-dimensional model using tracer breakthrough data and drawdown data simultaneously.

The exploratory analysis of hydraulic responses is carried out for further updating of the present three-dimensional hydro-structural model and to provide a starting point for the three-dimensional analysis of hydraulic responses which will be a test of the current structural model. The long-term pumping will provide the possibility of testing the current hydraulic conceptual model in more detail.The 3-D analysis uses drawdown and recovery curves in all observation sections simultaneously for parameter estimation. Features that may be deduced from this analysis include connectivity between the fracture zones surrounding the uranium mineralization (H1, V1, and those associated with the western and eastern granite), hydraulic transmissivity, storativity and hydraulic properties of boundaries. Depending on the current understanding of the system, alternative geometries may be tested.

One-dimensional analysis of tracer breakthrough can be justified since, in a radially converging tracer test, all transport may approximately be considered to take place along essentially one-dimensional transport paths. Evaluation of individual breakthrough curves will yield effective transport parameters for each analyzed path. The results are tracer residence times, dispersion coefficient (or alternatively Peclet numbers), and some measure of dilution. When using non-reactive tracers also the tracer mass recovery in the pumping borehole will yield valuable information about the physical nature of the flow paths.

If motivated by the results from the previous steps, a two-dimensional combined anlysis of hydraulic responses and solute transport will be made in the H1 fracture zone, where pressure drawdown, recovery, and tracer breakthrough curves are used simultaneously in parameter estimation. In a previous tracer experiment in a similar sub-horizontal fracture zone in Finnsjön, Sweden [7], tracer and pressure responses were evaluated separately. However, one conclusion was that the analysis would have benefitted from using both kind of data simultaneously for parameter estimation. Such concept is also supported by Medina et al. [8].

It is anticipated that the location and hydraulic properties of the boundaries will constitute the greatest difficulties for the intepretation of the flow and transport properties of the zone. In order to distinguish effects of heterogeneity/anisotopy within the zone from boundary effects, it is important that previous evaluation steps have already provided an adequate description of the boundaries to the zone, inluding vertical leakage. The expected results will be hydraulic transmissivity and storativity values, dispersivities, flow porosity, leakage parameters, and possibly boundary parameters. If data so indicates, hydraulic anisotropy values may also be estimated. Boundary parameters may be, for example, head values at constant head boundaries, or transmissivities of skin zones assigned along constant head boundaries.

The results of the evaluation of the combined pumping and tracer test will reveal properties of the natural flow system at the site. Such information is vital for a quantitative analysis of the forthcoming analoge transport study, which assesses the present mobilization and fracture controlled transport of radionuclides.

REFERENCES

[1] Blomqvist, R., Suksi, J., Ruskeeniemi, T., Ahonen, L., Niini, H, Vuorinen U., Jakobsson, K, 1995: The Palmottu Natural Analogue Project. Summary Report 1992-1994. Geological Survey of Finland, Report YST-88.

[2] Ruskeeniemi, T., Blomqvist, R., 1996: The Palmottu Natural Analogue Project. Progress Report 1995. Geological Survey of Finland, Report YST-93.

[3] Paananen, M., Blomqvist, R., 1994: The Palmottu Analogue Project. Conceptual Hydrogeological Model of Palmottu. Geological Survey of Finland, Report YST-86.

[4] Andersson, J-E., Ekman, L., Gustafsson, E., Nordqvist, R., Tirén, S., 1988: Hydraulic interference tests and tracer tests within the Brändan area, Finnsjön study site. The Fracture Zone Project - Phase 3. Swedish Nuclear Fuel and Waste Management Co (SKB). Technical Report 98-12.

[5] Gustafsson, E., Andersson, P. Ittner, T, Nordqvist, R., 1991: Large-scale Three-dimensional Tracer Test at Äspö. In Rhen I (ed) and Svensson U (ed) et al 1992. Äspö Hard Rock Laboratory: Evaluation of the combined long-term pumping and tracer test (LPT2) in borehole KAS06. Swedish Nuclear Fuel and Waste Management Co (SKB). Technical Report 92-32.

[6] García Gutiérrez, M., Yllera de Liano, A., Hernández Benitez, A., Guimerà, J., 1996: Results From the Field Tracer Tests in Boreholes S2 - S13 - S15 at El Berrocal Study Site, Spain. Draft report under the CIEMAT-ENRESA association agreement tracers tests contract C.A. Nº 702029. TR-14, EB-CIEMAT (96)2.

[7] Andersson, P., Nordqvist, R., Persson, T., Eriksson, C-O., Gustafsson, E., Ittner, T., 1993: Dipole Tracer Experiment in a Low-angle Fracture Zone at Finnsjön - Results and Interpretation. The Fracture Zone Project - Phase 3. Swedish Nuclear Fuel and Waste Management Co (SKB). Technical Report 93-26.

[8] Medina, A., Carrera, J., Galarza, G., 1990: Inverse modelling of coupled flow and solute transport problems. Modelcare 90 - Calibration and Reliability in groundwater modelling. IAHS Publ. No 195.

POSTER SESSION

The Assessment of Radionuclide Entrapment in Repository Host Rocks

W. Russell Alexander
GGWW, University of Berne, Switzerland

Bernhard Frieg
Nagra, Wettingen, Switzerland

Kunio Ota
PNC, Tono Geoscience Centre, Japan

Paul Bossart
Geotechnical Institute, Berne, Switzerland

Executive Summary

The Nagra/PNC field tracer migration experiment (MI) has just been completed following 10 years of intensive study in Nagra`s Grimsel Test Site (GTS) in central Switzerland (see Alexander et al., 1996a, elsewhere in this report). The Radionuclide Retardation Project (RRP) may be thought of taking over where MI left off but, this time, Nagra and PNC are examining specific retardation mechanisms of safety relevant radionuclides as they migrate through fractures in granite. The GTS MI experiment was specifically aimed at transport model testing and understanding how laboratory measured radionuclide retardation could be related to the real environment. In MI, several short-lived, weakly retarded, radionuclides were injected into a borehole in a water-bearing, complex fracture (or shear zone) in the GTS. The behaviour of the tracers was observed from other boreholes in the shear zone and the good fit between the model predictions, laboratory data and the field experimental observations was very encouraging.

The Radionuclide Retardation Project (RRP; see Alexander et al., 1996b) is now taking the MI work further in two main respects: first, strongly retarded radionuclides have been (ie September, 1996) injected, as in MI, into the same water conducting shear zone which has been used for the ten years work in MI and second, the sites of *in situ* radionuclide retardation will be physically defined (and not just assumed as in MI). The second part follows on partly from the first in that the strongly retarding radionuclides to be injected in RRP will travel through the experimental shear zone so slowly (some possibly taking years to decades compared with hours to months in MI) that it will be impossible to sit back and wait for their appearance at the extraction borehole as was the case in MI. Here, it will be necessary to physically excavate the entire experimental zone (approximately two tonnes of rock) and take sub-samples back to the laboratory to assess how far each radionuclide has travelled and compare these results with predictions based on laboratory experiments.

In this part of RRP, as well as producing data on retardation sites in the rock, a full 3D physical description of the shear zone will be carried out, so producing one of the most detailed descriptions of a water conducting shear zone in a rock to date. This should be of great use to transport modellers when constructing conceptual models of water flow systems in crystalline repository host rocks.

In addition, a second retardation mechanism is under study in RRP, namely matrix diffusion. This is where radionuclides diffuse out from the water conducting fractures and enter the actual matrix of the host rock via connected pores to be further retarded either by uptake on the rock pore surfaces or simply temporarily trapped in the pore waters. Work carried out on natural decay series disequilibrium several years ago showed that matrix diffusion is certainly occurring in the rock matrix close behind the experimental shear zone and so this seems a good area to define in detail the extent and form of *in situ* matrix diffusion. Specifically, the depth to which matrix porosity is connected to water conducting features such as the experimental shear zone will be studied as will the form of pore connectivity. This will provide information on the volumes of rock which could be available to retard radionuclides moving through crystalline repository host rocks and is intended to allow direct comparison with the large volume of laboratory experimental data on pore space availability and connectivity which probably over estimate both values due to experimental artefacts (such as pores opening when the rock samples are destressed on drilling *etc*).

Acknowledgements

Thanks to our numerous colleagues in the RRP and MI projects who have contributed to the success of the work over the last decade. Thanks also to Nagra and PNC for funding the projects.

References

Alexander et al., (1996a) GEOTRAP-FTTE Report, NEA OECD, Paris, France.

Alexander et al., (1996b) Grimsel Test Site 1996, Nagra Bulletin No 27, Nagra, Wettingen, Switzerland.

ANALYSIS OF UNDERGROUND NUCLEAR EXPLOSIONS IN ROCK SALT AT THE AZGIR SITE

E.B. Anderson, V.G. Savonenkov, Yu.M. Rogozin
RPA "V.G. Khlopin Radium Institute" St.Petersburg, Russia

L.R. Schneider, H. Krause
Stoller Ingenieurtechnik GmbH Dresden, Germany

In the early 1960s a programme of underground nuclear explosions used for peaceful purposes was developed in the Soviet Union. Further on, in 1965-1988 for scientific and national-economic purposes 128 charges were exploded at 115 technological sites. The comprehensive programme included investigations on the creation of underground reservoirs. The only geological medium retaining a stable cavity of explosion is rock salt. Industrio-experimental investigations of a technology for creation of underground cavities in rock salt were carried out in the south-western part of the Caspian Sea-side depression (at present it is Kazakhstan's territory) on the salt-dome elevation Great Azgir 270 km eastwards of Volgograd and 170 km northwards of Astrakhan (Fig. 1).

Fig. 1. The south-western part of the
Caspian Sea-side depression

The elevation is a complex fold of diapyric type. Its western part consists of a salt stock coming out to the surface and a closed salt dome. The eastern part is a closed dome extended in the submeridional direction. Positive salt structures are separated by a compensatory mould (depression) in whose centre rock salt beds at a depth of 2.0-2.5 km (Fig. 2).

The bedding depth of under-salt rocks is about 7 km. The total area of the Great Azgir structure is about 420 km². Deposits of rock salt formed in the low-Permian time 250 Mio. years ago. Over-salt rocks are represented by gyps and anhydrites of the low Permian, limestones and dolomites of the upper Permian, sandstones and aleurolites of the Permo-Trias, a thick formation of sandy-clayey rocks of the Neogene, sands and clays of sea sediments of the Quaternary time (Fig. 3).

All the over-salt rocks contain streaks and lenses saturated with aqueous solutions. The underground waters are slow or stagnant. By the chemical composition they are sodium-chloride solutions whose mineralization regularly increases with the depth up to 300 h/l.

To study the conditions for creation of large underground storage reservoirs a series of nuclear explosions were carried out in rocks of the Great Azgir salt-dome structure at 10 technological sites during 1966 -1979. All the underground explosions were conducted in vertical boreholes at depths assuring their complete containity. The hermetic sealing of the technological boreholes assured the radiation safety (Fig. 4).

1. Period of conducting the explosions:	22.04.1966 - 24.10.1979
2. Power energy release of the explosions :	1.1 kt - 100 kt (TNT)
3. Depth of loading the charges :	161 - 1 500 m
4. Free volume of the cavities :	10 000 - 240 000 m³

Fig. 4.   Underground nuclear explosions conducted in the Great Azgir structure

The conduction of the nuclear explosions was accompanied by comprehensive scientific-research programmes: geologo-hydrogeological, seismical, geochemical, radiochemical, nuclear-physical, ecological ones, etc.

The obtained information allowed to mark out two independent scientific lines:
- rock salt as a medium for isolation and disposal of radionuclides;
- underground nuclear explosions as large-scale geotechnical analogues of radioactive waste disposal.

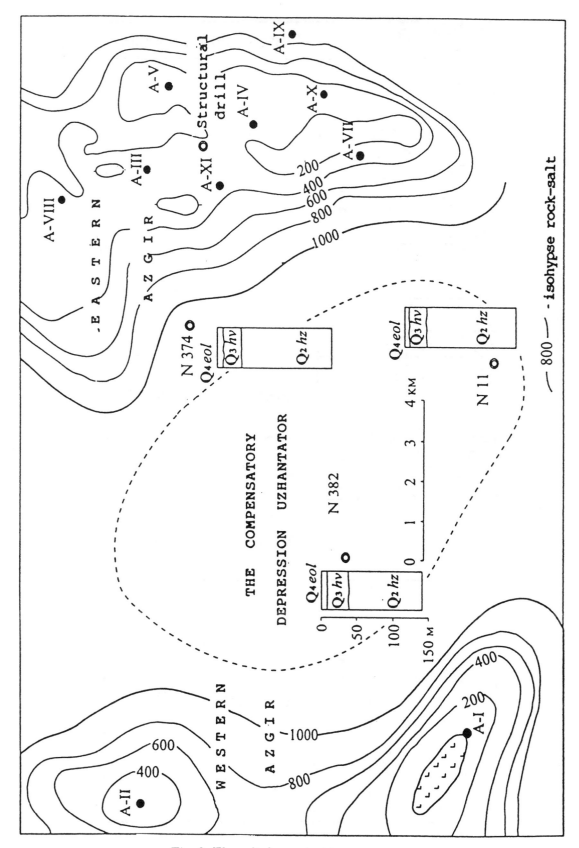

Fig. 2. The salt-dome structure
Great Azgir

221

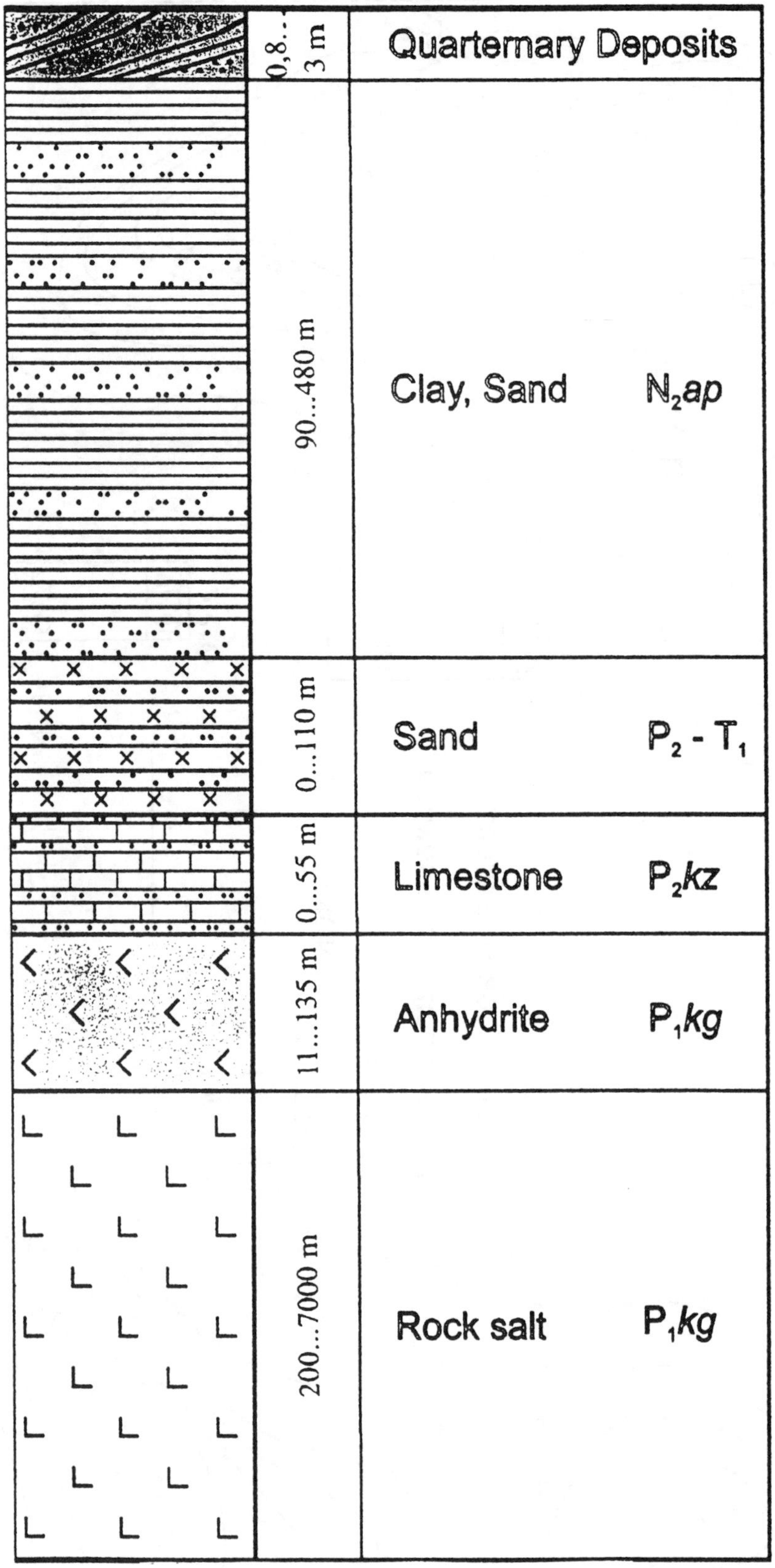

		Quarternary Deposits
	0,8...3 m	
	90...480 m	Clay, Sand N_2ap
	0...110 m	Sand $P_2 - T_1$
	0...55 m	Limestone P_2kz
	11...135 m	Anhydrite P_1kg
	200...7000 m	Rock salt P_1kg

Fig. 3. Over-salt deposits

A separate scientific programme provided prolonged observations of radionuclide migration processes and monitoring of environment objects: water bodies, soils, plants. Any underground nuclear explosion is accompanied by the incorporation and "burial" in geological formations of a large group of radionuclides: fission products, remaining fissile isotopes of uranium and plutonium, nuclides with induced activity. The different intensity of manifestation of an explosion's shock-thermal impact creates a certain zonality of the structure of its central and epicentral section. Three structural geochemical zones were discerned for nuclear explosions in rock salt: the interior, the boundary and the aureole ones. In Fig. 5 are shown the zones of a small power explosion conducted in the salt stock of the Western Azgir.

The interior zone is a lens of remolten rock salt on the bottom of the cavity. In an explosion 1 kt power about 1000 t of initially evaporated and molten salt are formed. It is characterized by the presence of specific "explosive" structures absent in natural salt rocks. Fig. 6 cites the concentrations of long-lived radionuclides: cesium-137 and strontium-90 in salts of the interior zones of three explosions.

Explosion	Concentration for the moment of formation, Bq/g	
	Cesium-137	Strontium-90
A-I	$1.4 \cdot 10^3$	$2.5 \cdot 10^3$
A-II	$1.5 \cdot 10^3$	
A-III	$8.0 \cdot 10^2 ... 1.5 \cdot 10^4$	$(1.7 ... 2.6) \cdot 10^2$

Fig. 6. The concentrations of cesium-137 and strontium-90 in remolten salts of the interior zone (for the moment of formation)

In the remolten salt lens there are present small quantities of neomorphic high-temperature minerals concentrating in their composition radionuclides of high-temperature elements, including isotopes of fissile materials.

The boundary zone is the boundary of the cavity, where the sharpest change-over of the substance's temperatures and phase state takes place. In this zone are concentrated radionuclides that are the most "sensitive" to the charge of physico-chemical conditions: antimony-125, ruthenium-106, etc.

The aureole radioactive zone is outside the cavity's limits and is represented by sections of shattered and cracked rocks. The maximum development of the zone proceeds over the cavity to distances 2.7- 4.4 of its radius. At the A-I explosion, however, individual cracks spread to greater distances, up to coming out to the surface of the salt stock. It was promoted by the steep, almost vertical slope of salt bed.

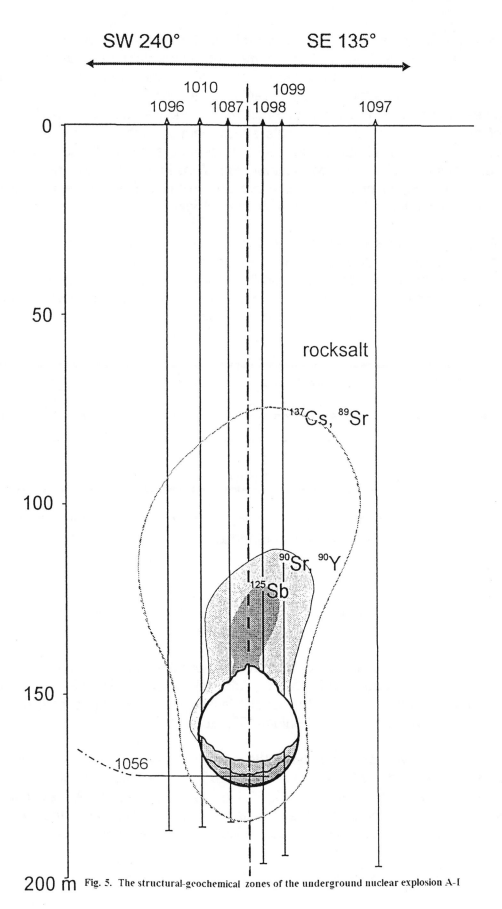

Fig. 5. The structural-geochemical zones of the underground nuclear explosion A-I

224

At an explosion the cracked zone is incorporated by gaseous and volatile radionuclides: radioactive noble gases that are predecessors of such nuclides as cesium-137 and strontium-90, isotopes of antimony and ruthenium. In the aureole zone there are 50-70% of the whole quantity of the formed cesium-137 and strontium-90. Their distribution over the zone is extremely uneven. They are present in form of easily soluble chlorides.

A study of products of nuclear explosions in rock salt showed, that the explosion cavity is a vast reaction space in which many chemical components interact in a broad range of temperatures and pressures. Multi-stage processes of mineral-formation take place, closely connected with the distribution of radionuclides. At explosions in rock salt two kinds of minerals are formed (Fig.7):
   - inherited, but fully transformed minerals: halite, anhydrite, calcite;
   - new minerals formed due to admixture elements: calcium, magnesium, silicon, iron, aluminium, as well as structural metals of the explosive device.

I group - inherited transformed minerals:
   ◇ halite $NaCl$;
   ◇ anhydrite $CaSO_4$;
   ◇ calcite $CaNO_3$.

II group - new minerals previously not present in rock salt:
   ◇ simple oxides - periclase MgO;
   ◇ complex oxides of magnetite group Fe_3O_4 (magnetite, maggemite, magnesioferrite, spinel);
   ◇ orthosilicates - monticellite $CaMgSiO_4$, forsterite $MgSiO_4$, bredigite Ca_2SiO_4;
   ◇ diorthosilicates - melilite-gelenite group $CaAl(Si_2O_7)$;
   ◇ simple sulfides - pyrite FeS_2.

Fig. 7. Main minerals (explosites) of underground nuclear explosions in rock salt.

The total mass of the new high-temperature minerals (or their slag formations) in relation to the total mass of remolten halite accounts for thousandths shares of per cent. These minerals and their slags, however, concentrate in their composition a large group of radionuclides of refractory elements: zirconium, cerium, europium, cobalt, uranium and plutonium. It is very important that these radionuclides' fixing in silicate-ferriferous neomorphs is sufficiently reliable, and the migration is unlikely. Regularities of the radionuclides' distribution by the mineral fractions of the A-IV explosion products are presented in Fig. 8.

CONCENTRATION

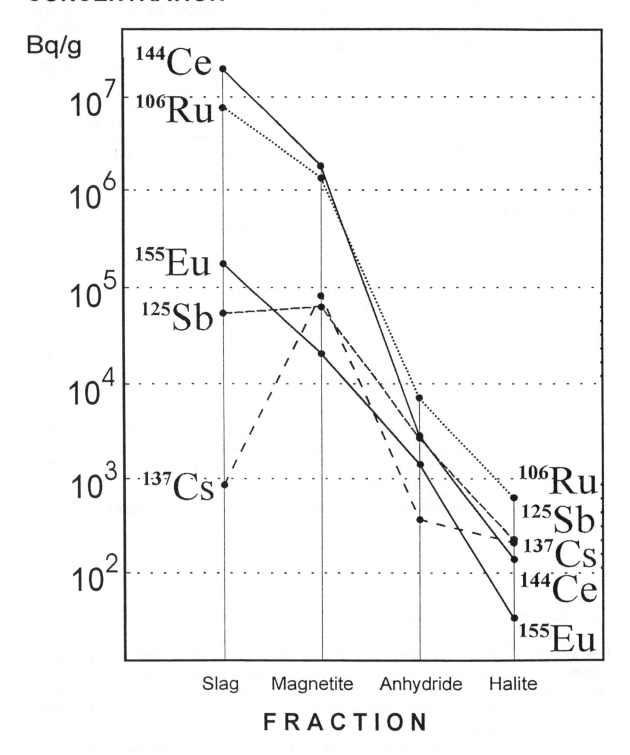

Fig. 8. The distribution of radionuclildes in the mineral fractions on the A-IV
explosion products (for the moment of formation)

The main mechanism of radionuclides migration in geological formations is their movement in the aqueous phase. The greatest radiation hazard for underground explosions in rock salt is represented by brines containing cesium-137 and strontium-90.

In the salt rocks of the Great Azgir 9 stable cavities were created with a total volume of about $1.2 \cdot 10^6$ m^3. At present five cavities, for different reasons and in different time, are filled with radioactive brine - a saturated solution of sodium chloride. The long-lived radionuclides cesium-137 and strontium-90 are present in the brines (Fig. 9).

Cavity of explosion	Volume of brine, m^3	Year of deter-mination	Concentration in brine, Bq/l	
			Cs-137	Sr-90
A-I	10 000	1966-1967	$2.4 \cdot 10^5$	$1.8 \cdot 10^4$
A-II	125 000	1968	$1.3 \cdot 10^5$	$0.3 \cdot 10^4$
A-III	215 000	1989	$3.4 \cdot 10^5$	$4.1 \cdot 10^4$
A-V	9 500	1980	$1.5 \cdot 10^5$	$3.0 \cdot 10^5$

Fig. 9. The concentration of long-lived radionuclides in the brines of nuclear explosion cavities

In sites of underground nuclear explosions migration channels of artificial origin arise: the aureole zone of crackness, splitting cracks in the over-salt rocks, finally, technological boreholes proper. Radionuclides can penetrate by these migration channels into the overlying water-bearing horizons and further to the environment.

For observation of the underground waters' radioactivity boreholes were drilled in the sites of the water-filled cavities at distances from 13 to 200 m from the technological borehole and from 50 to 280 m deep. In all the cases the waters analyzed were represented by sodium chloride solutions with a mineralization from 2.3 to 134 g/l.

A contamination of the brines was established in the subsurface part of the Western Azgir salt stock at the site of the explosion A-I (Fig. 10). The phenomenon is quite explicable, because in the vertically situated salt beds of the stock some explosion cracks came out to the surface under a narrow layer of sandy rocks and loams. An investigation of many boreholes, drilled at this site before and after the explosion, showed that the radius of the contamination aureole on the surface is about 300 m relative to the A-I borehole. A "technogenous" character of the radioactive contamination at the A-I site is not excluded either, because there for several years had been carried out experimental works for the extraction of great amounts of remolten radioactive salt from the cavity.

Index of boreholes	Distance from technological borehole, m	Depth, m	Concentrations, 1990-1991, Bq/l	
			Cs-137	Sr-90
A-I-1	13.5	50	161.5	14.8
A-I-2	25.0	50	141.5	-
A-I-3	49.0	50	18.6	3.0
A-II-1	30.0	225	0.04	0.05
A-II-2	70.0	225	0.03	0.02
A-III-1	200.0	250	0.66	0.37
A-IV-1	30.0	225	0.36	0.03
A-IV-2	70.0	225	0.08	0.02
A-V-1	50.0	280	6.1	0.15
A-V-2	100.0	280	17.7	3.7
Total activity of water in sources of water-supply			0.85	
Permissible concentrations in water, MPC			550	14.8

Fig. 10.   Concentrations of long-lived radionuclides in ground waters derived from data on observational boreholes in 1990-1991

Long-run observations were conducted at a site of explosion A-II. Analysis was carried out with the use of water samples taken from observational boreholes, sampling boreholes and technological borehole A-II. No contamination of water-bearing horizon was detected in Neogene sands, i.e. the nearest horizon to the roof of salt dome (Fig. 11).

In technological borehole A-II there is retained for some years a clear boundary of highly mineralized (300 g/l) dense radioactive brine to depth 300 m, less mineralized (30 g/l) and less radioactive (by 2 orders of magnitude) brine lying above.

The comparison of data obtained at different times for explosions A-III and A-IV (the Eastern Azgir) shows that the penetration of radionuclides into over-salt waters is most probably as a result of injection and soaking of radioactive rare gases at the explosion instant. Water-bearing horizons of the Eastern Azgir appear to be insulated, and waters have stagnant character. In 1980-1990 the concentration of cesium-137 varied in the range of 0.7-2.9 Bq/l according to the data on annual sampling. Conventional age of these waters determined by helium-argon ratio comes to $1.5-3.4 \cdot 10^6$ years.

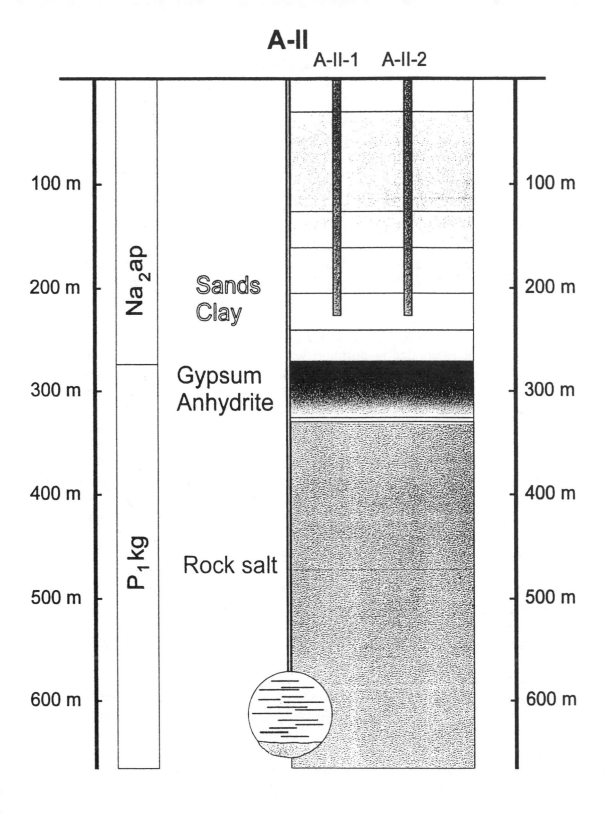

Fig. 11. Observational hydrogeological boreholes at the site of the explosion A-II

At present it is difficult to explain the mechanism of contamination of underground waters near the A-V explosion cavity (a water-bearing horizon in Neogene sands with a mineralization 135 g/l).

Analyzing the whole system of radionuclide migration in the Great Azgir salt-dome elevation, it is necessary to highlight its characteristic features (Fig. 12):

1) Cesium-137 and strontium-90 should be considered as the most migratable long-lived radionuclides. Long-lived isotopes of the fissile materials - plutonium and uranium - are quite strongly fixed in iron-silicate minerals in the bottom part of the cavity.

2) The migration channels are, first of all, disturbance zones and cracks formed at the explosion. All the cavities have connection with the surface through the technological borehole, and some of them, through observation and sampling boreholes. At boundaries of different by composition rocks at the explosion arise split cracks adjoining the technological borehole.

3) The migration medium is sodium-chloride solutions for which is characteristic:
    o A decrease of radionuclide sorption with an increase of mineralization. On sediments in observation boreholes, with a decrease of the solutions' mineralization from 117 down to 2.3 g/l, the sorption of cesium-137 increases: the distribution coefficient Kd increases from 16 to 1000.
    o For the Great Azgir, as well as for other regions of salt formations' development, is established the presence of gravitationally stable aquatic systems and a vertical "closedness" of salt deposits. Such a system includes several water-bearing horizons, where the sodium chloride mineralization increases as getting closer to the salt deposits. In the Great Azgir area from top to bottom by the gydrogeological section almost fresh waters are changed by saturated brines with a mineralization up to 256 g/l.

4) The geological environment and geochemical barriers.
    o The permeability of massive rock salt is extremely low: 10^{-9}-10^{-13} cm^2. At the same time the distribution coefficient for cesium and strontium in brines is extremely low, too, - less than a unity.
    o The Great Azgir salt rocks are overlapped (except the broken through stock Western Azgir) by a powerful thick of Neogene clays. Clayey rocks are water-proof and have a good sorption ability in relation to cesium, The Kd varies in saturated brines from 220 to 370.
    o Natural barriers for radioactive strontium are over-salt waters containing the sulfate ion. The initial ratio of this radionuclide to cesium-137 has sharply changed due to strontium sulfate precipitation in water-filled cavities of explosions.

5) The main mechanism of migration under the conditions of gravitationally stable aquatic systems is diffusion. The diffusion coefficient of cesium-137 and strontium-90 will not exceed 10^{-6} cm^2/s (tens of metres for hundreds of years).

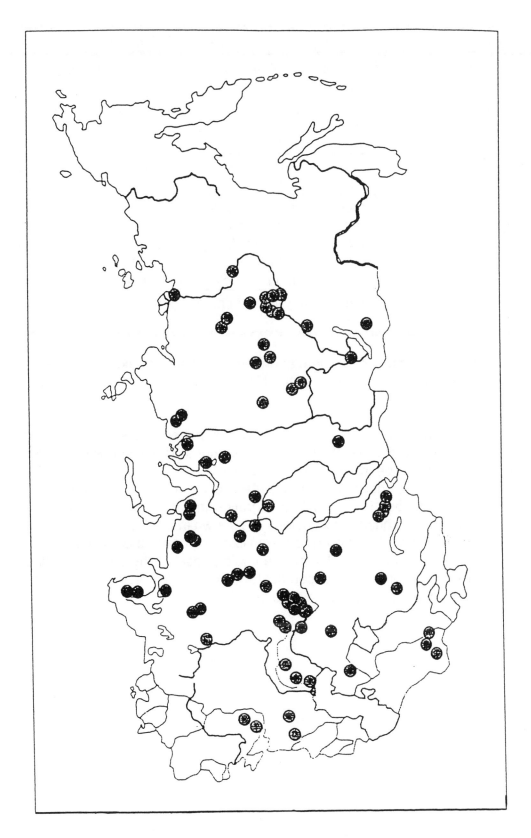

Fig. 13.   **Position of the 128 underground nuclear explosions at 115 test sites for peaceful scientifical and economical purposes (in salt, clay, tuff, granite)**

System components	Characteristic features
1. Radionuclides	Alkali and alkaline-earth metals: cesium-137 and strontium-90
2. Migration channels	Zones of disturbances and cracks formed by the explosion. Boreholes of various purposes. The rocks' natural porosity.
3. Migration medium	Sodium chloride solutions increasing the mineralization with the depth (2.3-256 g/l).
4. Geological environment and geochemical barriers	Rock salt with a low permeability in mass, with a low sorption of radionuclides. A thick of over-salt clayey rocks with a high sorption in relation to cesium (Kd = 220-370 in brines). Sulfate solution capable to precipitate strontium.
5. Main mechanism of migration	Diffusion (about 10^{-6} cm^2/s) - tens of metres for hundreds of years.

Fig. 12   Radionuclide migration system at underground nuclear explosion sites
             in the Great Azgir salt-dome structure

It should be noted, that migration is a complex multicomponent process that takes place at a simultaneous existence of several favourable conditions. The absence of at least one of the components may exclude the process on the whole.

The area of conduction of underground nuclear explosion, Great Azgir, is a unique object for studying the processes of radionuclide migration in salt formations. Other sites and host rocks of underground nuclear explosions are shown in fig. 13.

Here are in fact "buried", in different by stabilization forms, the radioactive waste of underground nuclear explosion, including the liquid phase. In the case of continuing the investigations, exclusively important experimental data may be obtained about the specific geotechnical "analogue" of radioactive waste burial.

Mechanical, Electronic and Instrumentation Development for Tracer Tests at El Berrocal Site.

M. García Gutiérrez; A. Hernández Benítez; A. Yllera de Llano; P. Rivas Romero
CIEMAT, Spain

J. Aleixandre; J. Bueno; O. González; J. Tamarit
CEDEX, Spain

J. Guimerà
UPC, Spain

Abstract

The international El Berrocal Project was an integrated exercise in geological, geochemical and hydrogeological characterization and had the aim of understanding and modelling the past and present-day migration processes that control the behaviour and distribution of naturally occurring radionuclides in a fractured granitic environment.

This paper contains the information concerning the desing and manufacture of the instrumentation used for the large scale tracer tests performed at El Berrocal.

The instrumentation development covers the following specifications:

• The measurement precission is equal or higher than 0.5% the full scale range.

• All the instrumentation chain corresponding to the sensors inserted in a borehole is controlled from the surface via a single 7 wire cable.

• The instrumentation cable runs along the whole length of the borehole and provides a measurement free of the electromagnetic noise from the inverters that activate the pumps.

• The data acquisition equipment is based on a microprocessor that allows the automatization of the measurements.The terminal for the measurement equipment is a portable PC that is also used to receive and analyze the essays. From it they can be defined the sensor parameters. It can also modify those, list them on screen or through a printer: From the PC the sensors can be also activated/deactivated, calibrated or followed on real time basis on the screen. The portable PC can be connected to any of the connexion boxes of all the modules. In any moment they can be seen the registered measurements for an essay, even if it is active. It can be also seen on screen the time evolution of the different physical magnitudes measured, as well as to obtain a print out of those measurements. The system admits up to 60 simultaneously active essays.

1. INTRODUCTION.

One of the main objectives of the large scale Tracer Test was the design and development of an instrumentation and data acquisition equipment; that would allow the automatic and precise measurement and record of all the physical parameters important for this type of essays. This approach has represented a great change concerning the clasical methodology used in this type of tests.

Three boreholes from El Berrocal experimental site have been instrumented for the large scale Tracer Test. Boreholes S-13 and S-15 for tracer injection and borehole S-2 for tracer recovery, Figure 1. For each borehole there are sensors to be measured by the surface equipment, associated to magnitudes registered in the tracer injection or recirculation circuits; and a second group of sensors placed in depth and associated to each one of the packered isolated sections in which the borehole has been physically divided.

For the performance of this test has been necessary an infraestructure that consists in a continous stabilized power supply and a deposit for formation water storage.

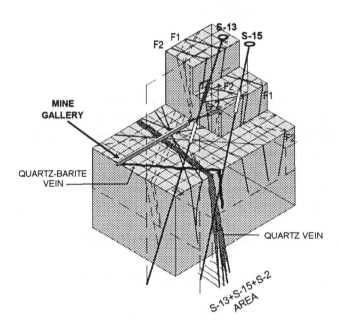

Figure 1. Location of the S-2, S-13 and S-15 boreholes at El Berrocal experiment site (Modified from R. Campos, CIEMAT 1993-1994).

2. MECHANICAL, ELECTRONIC AND INSTRUMENTATION DEVELOPMENT.

2.1 MECHANICAL DEVELOPMENT.

The most complicated borehole to instrument was S-13, that presents 5 water-tight separated sections, Figure 2. In each one is measured the pressure and temperature. In each of the injection sections, T2 and T4, is placed a pump associated to a power line and two tubes for water recirculation.

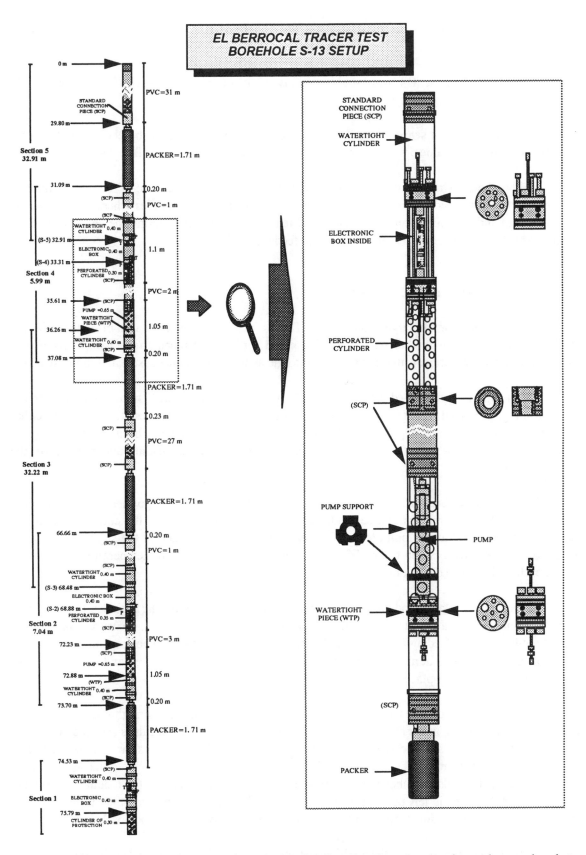

Figure 2. Scheme of the instrumentation used during the migration test: experimental setup placed at the S-13 injection borehole.

For the inflation of the packers are inserted two tubes, because they are serially connected in groups of two packers.

Borehole S-15 present 4 water-tight sections, with an injection interval in section T3. In borehole S-2 exist 2 water-tight sections.

The total number of sensors installed is 47; 22 of them inserted in the boreholes and the other 25 corresponding to the parameters for the magnitudes to be measured on the surface instrumentation.

The borehole instrumentation has required the design and manufacture of a great number of mechanical parts. It is especially important to point out the water-tight modules where are placed the electronic system and the transducers for presure and temperature, and that permit the connexion of the electronic cable and the cable for the pump electrical feeding.

2.2 ELECTRONIC DEVELOPMENT.

The electronic for the large scale Tracer Test is based in a modular design of the different parts that form the system. In that way, it is obtained a typical instrumentation chain formed by a transducer and signal conditioner, that is then routed to the analogical channel associated to the system.

The solution implemented has been a totally modular distributed system that through a universal routing, generated from the data acquisition equipment, allows the access to any sensor, either on the surface or placed at any depth inside the boreholes. Once the sensor has been routed the signals are sent through a current loop (it permits very long distances) for the reading and register by the measuring equipment.

The result have been a series of miniature electronic modules that go inside the water-tight boxes placed inside the boreholes. These modules include the control logical, the current/tension and tension/current converters and a multiplexor. To this module are also connected the signal conditioners designed to excite the sensors and to amplify the signal from the pressure and temperature transducers placed in each of the isolated sections.

The system is totally modular and it allows to connect serially all the modules that could be needed. It is only needed a seven wire cable for the routing (reset and clock), measurement (current loop) and module electric feeding (+/- 15 V. and ground). So, through the borehole opening it goes into the borehole a single cable that arrives to the first water-tight instrumentation box, and from there goes down serially to the all those placed below

For the read-out of a sensor it is send a Reset pulse, that sets all the multiplexors; from the serially connected modules, in the borehole pointing to the first sensor. Afterwards is sent the number of pulses required to select a specific sensor. It exists a logical control that counts the number of pulses that arrive and send to the multiplexor (with 16 channels) the code that corresponds to the sensor that is going to be read. If the pulse number is grater than 15, they are left to pass to the module inmediately below and the multiplexor is left pointing to the last channel; where is connected the output from the multiplexor placed in the module from the section located inmediately below in the borehole.

Once it has been selected the multiplexor channel to which is connected the output of the measured sensor signal converter, the signal in volts from the multiplexor output is converted to current to send it to the module inmediately above. There is converted back to tension to enter the multiplexor, and it follows this routine through all the modules until the signal reaches the microprocessor. The current is converted to tension and enters the digital/analogic converter from where it is read by the microprocessor and stored in memory.

The electronic modules permit to create a distributed measurement system and send the signals over long distances.

The existance of a microprocessor based measurement system allows to automatize the whole process and to let the system working autonomously once all the different essays programmed have been started.

It should be pointed out that in each electronic module two channels are reserved to connect the 5 V and ground references for that level. These channels are read in all measuring processes and the sensors readings are corrected with the values of those references.

This scheme allows to:

• Isolate the instrumentation from the measurement system.

• A universal coupling system between the instrumentation and the measurement system through a normalized interface.

• The coupling between instrumentation and measurement system, together with the real time measurement autocalibration, is what makes this system unique.

Essentially, the system works requesting the instrumentation a measurement from any of the user-defined transducers placed at the modules, on the surface or at depth, of the boreholes. For this, the sytem maps in its directions space requesting the instrumentation interface for a service. After being processed the result of this service request is the delivery of a measurement to the system.

The directions space defines the physical location of the transducer and the interface provides the access services to the instrumentation. The directions space is a three dimensional vector, where each direction defines a multiplexation section, level and order. The section refers to each one of the instrumented intervals, distinguishing between depth and surface for a same borehole. The level expresses each of the possible locations at depth of a transducer, and for that level the order selects a transducer chosen amongst all those available.

The real time measurement autocalibration has been thought to compensate, on one hand, the potential drifts between the signal reception points and the data acquisition by the system, the latter placed usually at a great distance. On the other hand, it normalizes through a standard the transfer function of the electronic chain. To do this two of the input channels to the multiplexor are used as calibration points for the measurement, feeding them with tension levels corresponding to both ends for the tension input range to the multiplexor. With those two points is obtained a calibration straight line used to displace the measurement and correct the potential drifts and the non normalization of the electronic components employed.

2.3 SURFACE INSTRUMENTATION.

The design of the new instrumentation has been mainly focused to develop a system that would present the following characteristics:

• Modular system: that could be extended or modified as necessary in a fast and easy way.

• That practically all the instrumentation could be also used in the laboratory, and that it could be easily replaced by any other model or commercial brand if necessary.

• That all the components could be used indistinctly in the injection or recovery zones, with the saving in spares and avoiding the need of duplicated material.

• That all the instruments have a 4-20mA analogic signal output or RS-232C serial port , to allow the continous record of the measured parameter.

• That all the instrumentation would be visible, to assure an easy performance and working conditions control. At the same time it has to be protected against extreme weather conditions and the dust and dirt present many times on long-term field experimentations.

The surface instrumentation is gathered mainly in two modules; one for the injection zone and the other for the recovery zone. They are two instrumentation cabinets, with inner sliding trays and on their sides have been mounted the injection and recovery panels respectively. The cabinets contain all the analytical instruments and electronic boxes to which all the sensors are connected. They also have a grounded power supply with enough capacity for all the instrumentation. They are sufficiently protected to resist long-term outdoors stays.

a) Injection zone instrumentation: The surface instrumentation designed takes into account the possibility of performing injections in three different zones. At the same time, it exists the need of carrying out homogeneization procedures and tracer dissapearance controls for the three zones. They have been built several basically identical panels, that allow the follow-up of an halide with the use of the proper ion selective electrodes or that of a dye through a colorimetric probe coupled to the recirculation circuit. In Figure 3 is shown the scheme of an injection panel. Panels are manufactured in plexiglass plastic material, and in them can be distinguished three zones that are differentiated by their function.

The lower zone contains the basic recirculation circuit. Inside the boreholes, in the lower part of each packered injection zone is placed a electrical pump that rises the water up to the panel passing it previously through a flowmeter: Once in the circuit the water is returned to the borehole after passing through a magnetic flowmeter. This flowmeter. placed at the end of the recirculation system, indicates both the instantaneous flow and the total cummulative volume recirculated. This basic circuit presents a sampling point, controlled by a valve, where it is possible to connect an automatic sampler or to perform a manual sampling. This valve is also used to blow off the air contained in the tubes, at the beginning of each pumping run. The circuit also has a pressure transducer and a one-way valve before the entrance of the flushing water coming from the surface deposit, for those cases of forced tracer injection.

The intermediate part of the panel corresponds to the analytical zone. Some of the pumped water can be circulated through it by the action of a valve connected to a potentiometer, for monitorization of the beginning of the analytical measurements. The water passes through a particle trap consisting in a level change where the particles are deposited. This avoids an inadequate weathering of the electrode membranes or a defective working of the ion selective electrodes used or the colorimeter probe located inmediately after the trap. The ionometer and the colorimeter are placed inside the cabinet on one of the sliding trays, making easier their access and the use of the instrument during the tests.

The upper zone is the one used for tracer injection. It contains an electric valve, that records the open/closed status, and a deposit where is placed the tracer that has been previously dissolved. This deposit is completely filled to avoid the entrance of air in the system. Of all the systems considered this one has been found to be the most safe and convenient. For the case that the tracer volume to inject would be greater than that of the deposit, this has an special cap to which it is possible the connection of other deposits. Another possibility for injection of large tracer volumes is to inject them through the deposit cleaning valve using a peristaltic pump.

The electrodes box, the one for the colorimetric probe, the particle trap and the injection deposit have been manufactured in transparent plexiglass that allows to observ the water passage through them.

In each injection panel is registered the moment at which begins the injection, the moment at which the analytical mesurements start, the analytical data (either the halide or dye concentration) and the instantaneous and total cummulative flows. It is also possible the sample collection, the clean-up of the tracer deposit and the injection of flushing water from the external deposit, being possible the register of the total and instantaneous flow.

b) Recovery zone instrumentation: The module placed at the extraction zone has two panels because there is a larger number of parameters to be measured, even with just one recovery zone at borehole S-2. In Figure 4 is shown the panel scheme for the recovery zone. The extraction module has mainly two zones, one for water recirculation and the other for analysis and control.

Due to the test characteristics and to the fact that in borehole S-2 water flows naturally at a flow of approximately 100 mL/min (flow that should be left to escape from the rock massif), it exists a regulating key at the exit of the recirculation circuit, after passing through a filter and a turbine flowmeter, to adjust that flow to the desired value. Due to the fact that water flows naturally to the recovery module, the recirculation is done with a peristaltic pump, that is working continuously to assure the homogeneization of the water that comes from the rock formation in the packered section at the bottom of borehole S-2. The recirculation flow and the amount of water involved are measured with a magnetic flowmeter located between the borehole and the cabinet. The recirculation zone has, as in the injection zone, a pressure transducer that is continuously registered and a valve for water sampling. Automatic samplers were placed in the water extraction zone.

The analytical zone is accessed via a valve with electric control, after which is placed a particle trap, and the two cells where are inserted the ion selective electrodes and the colorimetric probe: The circuit continues with another cell where are placed a pH and a conductivity probes; both of industrial type, that assure an efficient performance during all the duration of the tests.

The electronic boxes placed inside the recovery zone cabinet are connected to all the sensors that are continously monitorized. Amongst them are the pressure and temperature transducers for the instrumentation located inside the borehole, in the packered recovery section.

The injection and recovery zones, and the control laboratory have been continously interconnected. Although the follow-up of the essays is mainly done from the control center; it is also possible to use a portable computer as a terminal that can be connected at any of those points to know, in real time, the values of the measured parameters. It is also possible to change the type of essay, data collection frequency, recorded parameters, etc... If necessary, the data acquisition and control system allows to stablish new calibrations and/or controls opening and closing the data files where are stored the initial calibration conditions for each sensor.

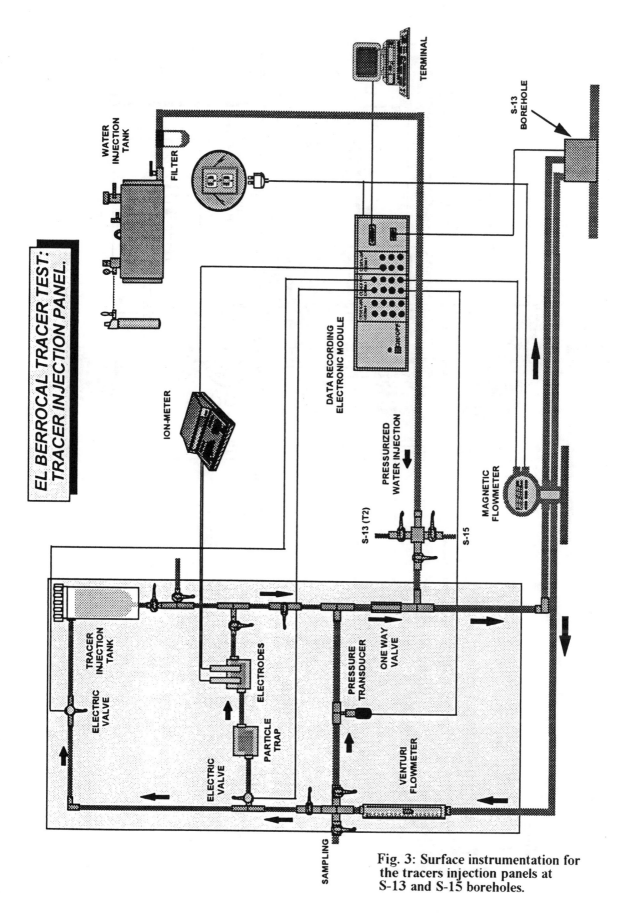

EL BERROCAL TRACER TEST: TRACER INJECTION PANEL.

WATER INJECTION TANK

FILTER

TERMINAL

S-13 BOREHOLE

ION-METER

DATA RECORDING ELECTRONIC MODULE

PRESSURIZED WATER INJECTION

S-13 (T2)

S-15

MAGNETIC FLOWMETER

TRACER INJECTION TANK

ELECTRIC VALVE

ELECTRODES

PARTICLE TRAP

ELECTRIC VALVE

PRESSURE TRANSDUCER

ONE WAY VALVE

VENTURI FLOWMETER

SAMPLING

Fig. 3: Surface instrumentation for the tracers injection panels at S-13 and S-15 boreholes.

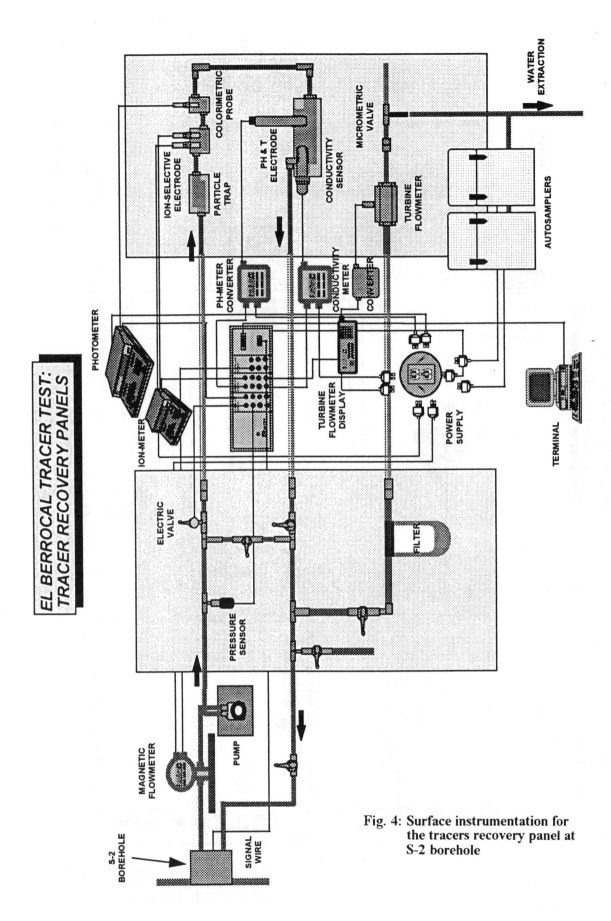

EL BERROCAL TRACER TEST: TRACER RECOVERY PANELS

Fig. 4: Surface instrumentation for the tracers recovery panel at S-2 borehole

Interpretation of Field Tracer Experiments Performed in Fractured Rock and Implications for a Performance Assessment

D Holton, M P Gardiner and N L Jefferies

(AEA Technology plc, 424.4, Harwell, Didcot, Oxon OX11 ORA, UK)

This abstract describes the interpretation of combined colloid and non-sorbing tracer migration tests performed in a convergent flow field in fractured rock. It describes the approach used and some of the difficulties encountered in the interpretation procedure. The work highlights the rationale of the interpretation within the broader context of the modelling of radionuclide transport for performance assessment purposes.

The potential for colloids to enhance radionuclide transport has been recognised in many radioactive waste disposal research and development programmes. The experiments described here have been undertaken to evaluate the mobility of different types of colloids in a fractured rock environment. The field experiments, undertaken at Reskageage Quarry Cornwall, are an extension of an earlier laboratory programme. The field site gives access to a longer pathway through a more complex fracture network than experienced in the well-controlled environment of the laboratory. Two types of colloids were investigated:

- monodisperse silica particles (30nm, 65nm, 800nm);

- monodisperse hematite(130nm).

The silica colloids were detected by photon correlation spectroscopy (PCS). Iron colloids, which tended to aggregate, were analysed using ICP-OES. Colloid migration was compared to that of a non-sorbing dye tracer (rhodamine-wt). Eight migration experiments (six silica colloid and two hematite colloid) were performed on length scales of 5m, 9.4m or 15.4m by the use of appropriate pairs of boreholes. Groundwater was abstracted from a borehole to achieve a radially-convergent flow field. Dye and colloid was released passively from neighbouring boreholes.

Break-through curves for two combined silica colloid and dye tracer tests described show a series of high-quality tracer break-through curves. The decay of tracers observed in the injection borehole is also measured. The main observations are:

- the velocities of silica colloids and rhodamine-wt are similar;

- relative to rhodamine-wt, transmission of 3Onm silica colloids ranges from 50% to 119%;

- results are highly reproducible.

The main observations from the hematite experiments are:

- hematite aggregates readily, but flocks are transported in the rapidly flowing water;

- transmission of hematite colloids is low, typically 10%.

The conceptual model for flow in fractured rock is a set of parallel fractures of uniform aperture through which water flows. Diffusion of solute into the rock matrix can occur. In addition there will be fractures, with negligible flow, connected to the flowing fractures, which will also provide paths for solute diffusion.

The interpretation approach involves fitting the observed normalised break-through with a semi-analytical method: inversion of the closed form Laplace transform solution. This method involves taking account of the mixing of solute in both the injection and pumped boreholes.

The long tail can be accounted for by RMD, but involves introducing a dense fracture spacing, which is inconsistent with the expected frequency of fractures allowing significant flow. Similar experiences have been found in the laboratory in which the tail can be accounted for by multiple flow paths.

The field tracer experiments, examples of which are described in this poster, can typically be used to:

- build confidence that the construction of a transport model from the gathered data is appropriate;

- build confidence in understanding transport processes;

- build confidence in a model by demonstrating the interpreted parameters are consistent with independent experiments and expert judgement.

The example tracer tests described in this poster were performed to build confidence in modelling processes in transport in a fractured rock. At a phenomenological level we found good agreement between the experimental observations and a relatively simple description of transport of solute. The break-through curves for the colloid showed little qualitative difference from the corresponding dye tracer experiment indicating significant mobility of the silica colloid.

A number of predictions were made about the likely results of a field experimental programme based on the results from the laboratory. In general, these predictions were confirmed. When, however, the results are examined in detail the added complexity makes the field experiments more difficult to interpret than the simple laboratory-based experiments. In the main, this difficulty no doubt stems from the complexities of the 'real' system. At a phenomenological level we found good agreement between the experimental observations and a relatively simple description of transport of solute.

This work has been funded by United Kingdom Nirex Limited as part of its Safety Assessment Research Programme (NSARP).

LIST OF PARTICIPANTS

BELGIUM

Geert VOLCKAERT
SCK-CEN
Boeretang 200
2400 Mol

Tel: +32 (14) 33 32 20
Fax: +32 (14) 32 35 53
Internet: gvolckae@sckcen.be

CANADA

Peter FLAVELLE
AECB
P.O. Box 1046, Station B
280 Slater Street
Ottawa, K18 5S9

Tel: +1 (613) 995 3816
Fax: +1 (613) 995 5086
Internet: flavelle.p@atomcon.gc.ca

Laurie FROST
AECL
Whiteshell Laboratories
Pinawa
Manitoba ROE 1L0

Tel: +1 (204) 753 2311
Fax: +1 (204) 345 8868
Internet: frostl@wl2.wl.aecl.ca

FINLAND

Runar BLOMQVIST
GTK
Betonimiehenkuja 4
02150 Espoo

Tel: +358 (20) 5 502469
Fax: +358 (20) 5 5012
Internet: runar.blomqvist@gsf.fi

Aimo HAUTOJÄRVI
VTT Energy
Tekniikantie 4 C, Espoo
P.O. Box 1604
02044 VTT

Tel: +358 (9) 456 5052
Fax: +358 (9) 456 5000
Internet: aimo.hautojarvi@vtt.fi

NOW AT:
Posiva Oy
Tel:+358 (9) 2280 3747
Fax:+358 (9) 2280 3719
Internet:aimo.hautojarvi@posiva.fi

Kai JAKOBSSON
STUK
P.O. Box 14
00881 Helsinki

Tel: +358 (9) 759 88 308
Fax: +358 (9) 759 88 382
Internet: kai.jakobsson@stuk.fi

Juhani VIRA
Posiva Oy
Mikonkatu 15 A
00100 Helsinki

Tel: +358 (9) 2280 3740
Fax: +358 (9) 2280 3719
Internet: juhani.vira@posiva.fi

FRANCE

Bernard BONIN
IPSN/DPRE/SERGD
CEN - FAR BP n° 6
92265 Fontenay-aux-roses Cédex

Tel: +33 (1) 46 54 73 96
Fax: +33 (1) 47 35 14 23
Internet: bonin@basilic.cea.fr

Lionel DEWIERE
ANDRA
Parc de la Croix Blanche
1-7, rue Jean Monnet
92298 Chatenay-Malabry Cédex

Tel: +33 (1) 46 11 80 38
Fax: +33 (1) 46 11 82 08
Internet: lionel.dewiere@andra.fr

GERMANY

Bruno BALTES
GRS
Schwertnergasse 1
50667 Köln

Tel: +49 (221) 2068 795
Fax: +49 (221) 2068 888
Internet: bat@ grs.de

Peter BOGORINSKI
GRS
Schwertnergasse 1
50667 Köln

Tel: +49 (221) 2068 809
Fax: +49 (221) 2068 888
Internet: bog@grs.de

Eckhard FEIN
GRS
Theodor-Heuss Strasse 4
38122 Braunschweig

Tel: +49 (531) 8012 292
Fax: +49 (531) 8012 200
Internet: fei@grs.de

Heidrun KRAUSE
Stoller Ingenieurtechnik GmbH
Schlüterstrasse 38
01277 Dresden

Tel: +49 (351) 3476209
Fax: +49 (351) 3476322
Internet: SIGDresden@aol.com

Klaus RÖHLIG
GRS
Schwertnergasse 1
50667 Köln

Tel: +49 (221) 2068 796
Fax: +49 (221) 2068 888
Internet: rkj@grs.de

Klaus SCHELKES
BGR
Stilleweg 2
30655 Hannover

Tel: +49 (511) 643 2616
Fax: +49 (511) 643 2304
Internet: Schelkes@gate1.bgr.d400.de

Lutz SCHNEIDER
Stoller Ingenieurtechnik GmbH
Schlüterstasse 38
01277 Dresden

Tel: +49 (351) 314 7213
Fax: +49 (351) 314 7222
Internet: SIGDresden@aol.com

Jürgen WOLLRATH
BfS
P.O. Box 10 01 49
38201 Salzgitter

Tel: +49 (531) 592 7704
Fax: +49 (531) 592 7614
Internet: jwollrath@bfs.de

JAPAN

Masahiro UCHIDA
PNC
Tokai Works
Muramatsu, Tokai-Mura
Ibaraki, 319-11

Tel: +81 (29) 287 0928
Fax: +81 (29) 287 3258
Internet: uchida@tokai.pnc.go.jp

KOREA

Chul-Hyung KANG
KAERI/NEMAC
P.O. Box 105, Yusong
Taejon, 305-600

Tel: +82 (42) 868 2853
Fax: +82 (42) 868 8850
Internet: chkang@kaeri.re.kr

RUSSIA

Vladimir SAVONENKOV
V.G. Khlopin Radium Institute
2-nd Murinsky Ave., 28
194021 St. Petersburg

SPAIN

Jesus ALONSO
Enresa
Emilio Vargas 7
28043 Madrid

Tel: +34 (1) 566 8100
Fax: +34 (1) 566 8165

Julio ASTUDILLO
Enresa
Emilio Vargas 7
28043 Madrid

Tel: +34 (1) 566 8120
Fax: +34 (1) 566 8165
Internet: jasp@enresa.es

Carmen BAJOS PARADA
Enresa
Emilio Vargas 7
28043 Madrid

Tel: +34 (1) 566 8100
Fax: +34 (1) 566 8165

Jordi GUIMERA
Dep. Enginyeria del Terreny
Civil Engineering School
UPC
Campus nord
C/.Gran Capita s/n, Ed. D-2
08034 - Barcelona

Tel: +34 (3) 401 7259/4016890
Fax: +34 (3) 401 6504
Internet: guimera@etseccpb.upc.es

SWEDEN

Peter ANDERSSON
Geosigma AB
Box 894
75108 Uppsala

Tel: +46 (18) 65 08 15
Fax: +46 (18) 12 13 02
Internet:
peter.andersson@geosigma.se

Erik GUSTAFSSON
Geosigma AB
Box 894
75108 Uppsala

Tel: +46 (18) 65 08 14
Fax: +46 (18) 12 13 02
Internet:
erik.gustafsson@geosigma.se

Rune NORDQVIST
Geosigma AB
Box 894
75108 Uppsala

Tel: +46 (18) 65 0826
Fax: +46 (18) 12 1302
Internet:
rune.nordqvist@geosigma.se

Olle OLSSON
SKB Äspö HRL
PL 300
572 95 Figeholm

Tel: +46 (491) 76 78 09
Fax: +46 (491) 82 00 5
Internet: skboo@skb.se

Anders WINBERG
Conterra AB
Ögärdesvägen 4
433 30 Parthle

Tel: +46 (31) 44 32 90
Fax: +46 (31) 44 32 89
Internet:
anders.winberg@mailbox.swipnet.se

SWITZERLAND

Russell ALEXANDER
Nagra
Hardstrasse 73
5430 Wettingen

Tel: +41 (56) 4371 320
Fax: +41 (56) 4371 317
Internet: russell@mpi.unibe.ch

and

GGWW, University of Berne
Baltzerstrasse 1
3012 Berne

Tel: +41 (31) 631 8520
Fax: +41 (31) 631 4843
Internet: russell@mpi.unibe.ch

Walter HEER
PSI
OFLA/208
5232 Villigen PSI

Tel: +41 (56) 310 4114
Fax: +41 (56) 310 2821
Internet: walter.heer@psi.ch

Andreas JAKOB
PSI
OFLA/203
5232 Villigen PSI

Tel: +41 (56) 310 2420
Fax: +41 (56) 310 2821
Internet: andreas.jakob@psi.ch

Johannes VIGFUSSON
HSK
5232 Villigen HSK

Tel: +41 (56) 310 3974
Fax: +41 (56) 310 3907
Internet: vj@hskib4.hsk.psi.ch

UNITED KINGDOM

Mike HEATH
Earth Resources Centre
University of Exeter
Laver Building - North Park Road
Exeter, UK EX4 4QE

Tel: +44 (1209) 216 647
Fax: +44 (1209) 216 647

David HOLTON
AEA Technology
424.4 Harwell
Didcot, Oxon, OX11 0RA

Tel: +44 (1235) 436 763
Fax: +44 (1235) 436 579
Internet: david.holton@aeat.co.uk

Paul SMITH
Safety Assessment Management Ltd
10 the Paddock
Bingham
Nottinghamshire NG13 8HQ

Tel: +44 (1949) 836 808
Fax: as above
Internet: paul@samltd.demon.co.uk

USA

Richard BEAUHEIM
Sandia National Laboratories
PO BOX 5800, Mail Stop 1324
Organization 6115
Albuquerque, New Mexico
87185-1324

Tel: +1 (505) 848 0675
Fax: +1 (505) 848 0605
Internet: rlbeauh@nwer.sandia.gov

Richard CODELL
US NRC
Washington D.C. 20555

Tel: +1 (301) 415 8167
Fax: +1 (301) 415 5399
Internet: rbc@nrc.gov

Thomas W. DOE
Golder Associates Inc.
4104-148th avenue N.E.
Redmond, Washington 98052

Tel: +1 (206) 883 0777
Fax: +1 (206) 882 5498
Internet: tdoe@golder.com

Lucy MEIGS
Sandia National Laboratories
PO BOX 5800, Mail Stop 1324
Organization 6115
Albuquerque, New Mexico
87185-1324

Tel: +1 (505) 848 0507
Fax: +1 (505) 848 0605
Internet: lcmeigs@nwer.sandia.gov

EC

Henning von MARAVIC
European Commission
EC-DG12-F5
Rue de la Loi 200, T-61, E-S 4
1049 Brussels
Belgium

Tel: +32 2 29 65 273
Fax: +32 2 29 66883/54991
Internet:
henning.ritter-von-maravic
@dg12.cec.be

NEA

Philippe LALIEUX
OECD Nuclear Energy Agency
Le Seine St-Germain
12 boulevard des Îles
92130 Issy-les-Moulineaux
France

Tel: +33 (1) 45 24 10 47
Fax: +33 (1) 45 24 11 10
Internet: lalieux@nea.fr

Claudio PESCATORE
OECD Nuclear Energy Agency
Le Seine St-Germain
12 boulevard des Îles
92130 Issy-les-Moulineaux
France

Tel: +33 (1) 45 24 11 04
Fax: +33 (1) 45 24 11 10
Internet: pescatore@nea.fr

MAIN SALES OUTLETS OF OECD PUBLICATIONS
PRINCIPAUX POINTS DE VENTE DES PUBLICATIONS DE L'OCDE

AUSTRALIA – AUSTRALIE
D.A. Information Services
648 Whitehorse Road, P.O.B 163
Mitcham, Victoria 3132   Tel. (03) 9210.7777
Fax: (03) 9210.7788

AUSTRIA – AUTRICHE
Gerold & Co.
Graben 31
Wien I   Tel. (0222) 533.50.14
Fax: (0222) 512.47.31.29

BELGIUM – BELGIQUE
Jean De Lannoy
Avenue du Roi, Koningslaan 202
B-1060 Bruxelles   Tel. (02) 538.51.69/538.08.41
Fax: (02) 538.08.41

CANADA
Renouf Publishing Company Ltd.
5369 Canotek Road
Unit 1
Ottawa, Ont. K1J 9J3   Tel. (613) 745.2665
Fax: (613) 745.7660

Stores:
71 1/2 Sparks Street
Ottawa, Ont. K1P 5R1   Tel. (613) 238.8985
Fax: (613) 238.6041

12 Adelaide Street West
Toronto, QN M5H 1L6   Tel. (416) 363.3171
Fax: (416) 363.5963

Les Éditions La Liberté Inc.
3020 Chemin Sainte-Foy
Sainte-Foy, PQ G1X 3V6   Tel. (418) 658.3763
Fax: (418) 658.3763

Federal Publications Inc.
165 University Avenue, Suite 701
Toronto, ON M5H 3B8   Tel. (416) 860.1611
Fax: (416) 860.1608

Les Publications Fédérales
1185 Université
Montréal, QC H3B 3A7   Tel. (514) 954.1633
Fax: (514) 954.1635

CHINA – CHINE
Book Dept., China National Publications
Import and Export Corporation (CNPIEC)
16 Gongti E. Road, Chaoyang District
Beijing 100020   Tel. (10) 6506-6688 Ext. 8402
(10) 6506-3101

CHINESE TAIPEI – TAIPEI CHINOIS
Good Faith Worldwide Int'l. Co. Ltd.
9th Floor, No. 118, Sec. 2
Chung Hsiao E. Road
Taipei   Tel. (02) 391.7396/391.7397
Fax: (02) 394.9176

**CZECH REPUBLIC –
RÉPUBLIQUE TCHÈQUE**
National Information Centre
NIS – prodejna
Konviktská 5
Praha 1 – 113 57   Tel. (02) 24.23.09.07
Fax: (02) 24.22.94.33
E-mail: nkposp@dec.niz.cz
Internet: http://www.nis.cz

DENMARK – DANEMARK
Munksgaard Book and Subscription Service
35, Nørre Søgade, P.O. Box 2148
DK-1016 København K   Tel. (33) 12.85.70
Fax: (33) 12.93.87

J. H. Schultz Information A/S,
Herstedvang 12,
DK – 2620 Albertslung   Tel. 43 63 23 00
Fax: 43 63 19 69
Internet: s-info@inet.uni-c.dk

EGYPT – ÉGYPTE
The Middle East Observer
41 Sherif Street
Cairo   Tel. (2) 392.6919
Fax: (2) 360.6804

FINLAND – FINLANDE
Akateeminen Kirjakauppa
Keskuskatu 1, P.O. Box 128
00100 Helsinki

Subscription Services/Agence d'abonnements :
P.O. Box 23
00100 Helsinki   Tel. (358) 9.121.4403
Fax: (358) 9.121.4450

*FRANCE
OECD/OCDE
Mail Orders/Commandes par correspondance :
2, rue André-Pascal
75775 Paris Cedex 16   Tel. 33 (0)1.45.24.82.00
Fax: 33 (0)1.49.10.42.76
Telex: 640048 OCDE
Internet: Compte.PUBSINQ@oecd.org

Orders via Minitel, France only/
Commandes par Minitel, France exclusivement :
36 15 OCDE

OECD Bookshop/Librairie de l'OCDE :
33, rue Octave-Feuillet
75016 Paris   Tel. 33 (0)1.45.24.81.81
33 (0)1.45.24.81.67

Dawson
B.P. 40
91121 Palaiseau Cedex   Tel. 01.89.10.47.00
Fax: 01.64.54.83.26

Documentation Française
29, quai Voltaire
75007 Paris   Tel. 01.40.15.70.00

Economica
49, rue Héricart
75015 Paris   Tel. 01.45.78.12.92
Fax: 01.45.75.05.67

Gibert Jeune (Droit-Économie)
6, place Saint-Michel
75006 Paris   Tel. 01.43.25.91.19

Librairie du Commerce International
10, avenue d'Iéna
75016 Paris   Tel. 01.40.73.34.60

Librairie Dunod
Université Paris-Dauphine
Place du Maréchal-de-Lattre-de-Tassigny
75016 Paris   Tel. 01.44.05.40.13

Librairie Lavoisier
11, rue Lavoisier
75008 Paris   Tel. 01.42.65.39.95

Librairie des Sciences Politiques
30, rue Saint-Guillaume
75007 Paris   Tel. 01.45.48.36.02

P.U.F.
49, boulevard Saint-Michel
75005 Paris   Tel. 01.43.25.83.40

Librairie de l'Université
12a, rue Nazareth
13100 Aix-en-Provence   Tel. 04.42.26.18.08

Documentation Française
165, rue Garibaldi
69003 Lyon   Tel. 04.78.63.32.23

Librairie Decitre
29, place Bellecour
69002 Lyon   Tel. 04.72.40.54.54

Librairie Sauramps
Le Triangle
34967 Montpellier Cedex 2   Tel. 04.67.58.85.15
Fax: 04.67.58.27.36

A la Sorbonne Actual
23, rue de l'Hôtel-des-Postes
06000 Nice   Tel. 04.93.13.77.75
Fax: 04.93.80.75.69

GERMANY – ALLEMAGNE
OECD Bonn Centre
August-Bebel-Allee 6
D-53175 Bonn   Tel. (0228) 959.120
Fax: (0228) 959.12.17

GREECE – GRÈCE
Librairie Kauffmann
Stadiou 28
10564 Athens   Tel. (01) 32.55.321
Fax: (01) 32.30.320

HONG-KONG
Swindon Book Co. Ltd.
Astoria Bldg. 3F
34 Ashley Road, Tsimshatsui
Kowloon, Hong Kong   Tel. 2376.2062
Fax: 2376.0685

HUNGARY – HONGRIE
Euro Info Service
Margitsziget, Európa Ház
1138 Budapest   Tel. (1) 111.60.61
Fax: (1) 302.50.35
E-mail: euroinfo@mail.matav.hu
Internet: http://www.euroinfo.hu//index.html

ICELAND – ISLANDE
Mál og Menning
Laugavegi 18, Pósthólf 392
121 Reykjavik   Tel. (1) 552.4240
Fax: (1) 562.3523

INDIA – INDE
Oxford Book and Stationery Co.
Scindia House
New Delhi 110001   Tel. (11) 331.5896/5308
Fax: (11) 332.2639
E-mail: oxford.publ@axcess.net.in

17 Park Street
Calcutta 700016   Tel. 240832

INDONESIA – INDONÉSIE
Pdii-Lipi
P.O. Box 4298
Jakarta 12042   Tel. (21) 573.34.67
Fax: (21) 573.34.67

IRELAND – IRLANDE
Government Supplies Agency
Publications Section
4/5 Harcourt Road
Dublin 2   Tel. 661.31.11
Fax: 475.27.60

ISRAEL – ISRAËL
Praedicta
5 Shatner Street
P.O. Box 34030
Jerusalem 91430   Tel. (2) 652.84.90/1/2
Fax: (2) 652.84.93

R.O.Y. International
P.O. Box 13056
Tel Aviv 61130   Tel. (3) 546 1423
Fax: (3) 546 1442
E-mail: royil@netvision.net.il

Palestinian Authority/Middle East:
INDEX Information Services
P.O.B. 19502
Jerusalem   Tel. (2) 627.16.34
Fax: (2) 627.12.19

ITALY – ITALIE
Libreria Commissionaria Sansoni
Via Duca di Calabria, 1/1
50125 Firenze   Tel. (055) 64.54.15
Fax: (055) 64.12.57
E-mail: licosa@ftbcc.it

Via Bartolini 29
20155 Milano   Tel. (02) 36.50.83

Editrice e Libreria Herder
Piazza Montecitorio 120
00186 Roma   Tel. 679.46.28
Fax: 678.47.51

Libreria Hoepli
Via Hoepli 5
20121 Milano   Tel. (02) 86.54.46
Fax: (02) 805.28.86

Libreria Scientifica
Dott. Lucio de Biasio 'Aeiou'
Via Coronelli, 6
20146 Milano   Tel. (02) 48.95.45.52
                      Fax: (02) 48.95.45.48

JAPAN – JAPON
OECD Tokyo Centre
Landic Akasaka Building
2-3-4 Akasaka, Minato-ku
Tokyo 107   Tel. (81.3) 3586.2016
                      Fax: (81.3) 3584.7929

KOREA – CORÉE
Kyobo Book Centre Co. Ltd.
P.O. Box 1658, Kwang Hwa Moon
Seoul   Tel. 730.78.91
                      Fax: 735.00.30

MALAYSIA – MALAISIE
University of Malaya Bookshop
University of Malaya
P.O. Box 1127, Jalan Pantai Baru
59700 Kuala Lumpur
Malaysia   Tel. 756.5000/756.5425
                      Fax: 756.3246

MEXICO – MEXIQUE
OECD Mexico Centre
Edificio INFOTEC
Av. San Fernando no. 37
Col. Toriello Guerra
Tlalpan C.P. 14050
Mexico D.F.   Tel. (525) 528.10.38
                      Fax: (525) 606.13.07

E-mail: ocde@rtn.net.mx

NETHERLANDS – PAYS-BAS
SDU Uitgeverij Plantijnstraat
Externe Fondsen
Postbus 20014
2500 EA's-Gravenhage Tel. (070) 37.89.880
Voor bestellingen: Fax: (070) 34.75.778

Subscription Agency/   Agence d'abonnements :
SWETS & ZEITLINGER BV
Heereweg 347B
P.O. Box 830
2160 SZ Lisse   Tel. 252.435.111
                      Fax: 252.415.888

**NEW ZEALAND –
NOUVELLE-ZÉLANDE**
GPLegislation Services
P.O. Box 12418
Thorndon, Wellington Tel. (04) 496.5655
                      Fax: (04) 496.5698

NORWAY – NORVÈGE
NIC INFO A/S
Ostensjoveien 18
P.O. Box 6512 Etterstad
0606 Oslo   Tel. (22) 97.45.00
                      Fax: (22) 97.45.45

PAKISTAN
Mirza Book Agency
65 Shahrah Quaid-E-Azam
Lahore 54000   Tel. (42) 735.36.01
                      Fax: (42) 576.37.14

PHILIPPINE – PHILIPPINES
International Booksource Center Inc.
Rm 179/920 Cityland 10 Condo Tower 2
HV dela Costa Ext cor Valero St.
Makati Metro Manila   Tel. (632) 817 9676
                      Fax: (632) 817 1741

POLAND – POLOGNE
Ars Polona
00-950 Warszawa
Krakowskie Prezdmiescie 7   Tel. (22) 264760
                            Fax: (22) 265334

PORTUGAL
Livraria Portugal
Rua do Carmo 70-74
Apart. 2681
1200 Lisboa   Tel. (01) 347.49.82/5
                      Fax: (01) 347.02.64

SINGAPORE – SINGAPOUR
Ashgate Publishing
Asia Pacific Pte. Ltd
Golden Wheel Building, 04-03
41, Kallang Pudding Road
Singapore 349316   Tel. 741.5166
                      Fax: 742.9356

SPAIN – ESPAGNE
Mundi-Prensa Libros S.A.
Castelló 37, Apartado 1223
Madrid 28001   Tel. (91) 431.33.99
                      Fax: (91) 575.39.98
E-mail: mundiprensa@tsai.es
Internet: http://www.mundiprensa.es

Mundi-Prensa Barcelona
Consell de Cent No. 391
08009 – Barcelona   Tel. (93) 488.34.92
                      Fax: (93) 487.76.59

Libreria de la Generalitat
Palau Moja
Rambla dels Estudis, 118
08002 – Barcelona
           (Suscripciones) Tel. (93) 318.80.12
           (Publicaciones) Tel. (93) 302.67.23
                           Fax: (93) 412.18.54

SRI LANKA
Centre for Policy Research
c/o Colombo Agencies Ltd.
No. 300-304, Galle Road
Colombo 3   Tel. (1) 574240, 573551-2
                      Fax: (1) 575394, 510711

SWEDEN – SUÈDE
CE Fritzes AB
S-106 47 Stockholm   Tel. (08) 690.90.90
                      Fax: (08) 20.50.21

For electronic publications only/
Publications électroniques seulement
STATISTICS SWEDEN
Informationsservice
S-115 81 Stockholm   Tel. 8 783 5066
                      Fax: 8 783 4045

Subscription Agency/Agence d'abonnements :
Wennergren-Williams Info AB
P.O. Box 1305
171 25 Solna   Tel. (08) 705.97.50
                      Fax: (08) 27.00.71

Liber distribution
Internatinal organizations
Fagerstagatan 21
S-163 52 Spanga

SWITZERLAND – SUISSE
Maditec S.A. (Books and Periodicals/Livres
et périodiques)
Chemin des Palettes 4
Case postale 266
1020 Renens VD 1   Tel. (021) 635.08.65
                      Fax: (021) 635.07.80

Librairie Payot S.A.
4, place Pépinet
CP 3212
1002 Lausanne   Tel. (021) 320.25.11
                      Fax: (021) 320.25.14

Librairie Unilivres
6, rue de Candolle
1205 Genève   Tel. (022) 320.26.23
                      Fax: (022) 329.73.18

Subscription Agency/Agence d'abonnements :
Dynapresse Marketing S.A.
38, avenue Vibert
1227 Carouge   Tel. (022) 308.08.70
                      Fax: (022) 308.07.99

See also – Voir aussi :
OECD Bonn Centre
August-Bebel-Allee 6
D-53175 Bonn (Germany)   Tel. (0228) 959.120
                         Fax: (0228) 959.12.17

THAILAND – THAÏLANDE
Suksit Siam Co. Ltd.
113, 115 Fuang Nakhon Rd.
Opp. Wat Rajbopith
Bangkok 10200   Tel. (662) 225.9531/2
                      Fax: (662) 222.5188

**TRINIDAD & TOBAGO, CARIBBEAN
TRINITÉ-ET-TOBAGO, CARAÏBES**
Systematics Studies Limited
9 Watts Street
Curepe
Trinidad & Tobago, W.I.   Tel. (1809) 645.3475
                          Fax: (1809) 662.5654
E-mail: tobe@trinidad.net

TUNISIA – TUNISIE
Grande Librairie Spécialisée
Fendri Ali
Avenue Haffouz Imm El-Intilaka
Bloc B 1 Sfax 3000   Tel. (216-4) 296 855
                      Fax: (216-4) 298.270

TURKEY – TURQUIE
Kültür Yayinlari Is-Türk Ltd.
Atatürk Bulvari No. 191/Kat 13
06684 Kavaklidere/Ankara
                      Tel. (312) 428.11.40 Ext. 2458
                      Fax : (312) 417.24.90

Dolmabahce Cad. No. 29
Besiktas/Istanbul   Tel. (212) 260 7188

UNITED KINGDOM – ROYAUME-UNI
The Stationery Office Ltd.
Postal orders only:
P.O. Box 276, London SW8 5DT
Gen. enquiries   Tel. (171) 873 0011
                      Fax: (171) 873 8463

The Stationery Office Ltd.
Postal orders only:
49 High Holborn, London WC1V 6HB
Branches at: Belfast, Birmingham, Bristol,
Edinburgh, Manchester

UNITED STATES – ÉTATS-UNIS
OECD Washington Center
2001 L Street N.W., Suite 650
Washington, D.C. 20036-4922 Tel. (202) 785.6323
                            Fax: (202) 785.0350
Internet: washcont@oecd.org

Subscriptions to OECD periodicals may also be
placed through main subscription agencies.

Les abonnements aux publications périodiques de
l'OCDE peuvent être souscrits auprès des
principales agences d'abonnement.

Orders and inquiries from countries where Distribu-
tors have not yet been appointed should be sent to:
OECD Publications, 2, rue André-Pascal, 75775
Paris Cedex 16, France.

Les commandes provenant de pays où l'OCDE n'a
pas encore désigné de distributeur peuvent être
adressées aux Éditions de l'OCDE, 2, rue André-
Pascal, 75775 Paris Cedex 16, France.

12-1996

OECD PUBLICATIONS, 2, rue André-Pascal, 75775 PARIS CEDEX 16
PRINTED IN FRANCE
(66 97 13 1 P) ISBN 92-64-16013-2 – No. 49883 1997